D1574139

A Study of

A Prototype Floating Community

by

TRITON FOUNDATION, INC.

R. Buckminster Fuller

for the

US Department of Housing and Urban Development

University Press of the Pacific
Honolulu, Hawaii

A Study of a Prototype Floating Community

by
Triton Foundation
R. Buckminster Fuller

for the
U. S. Department of Housing and Urban Development

Afterword by Adam Starchild

ISBN: 1-4102-1818-X

Reprinted from the 1968 edition

University Press of the Pacific
Honolulu, Hawaii
http://www.universitypressofthepacific.com

TRITON CITY

A PROTOTYPE FLOATING COMMUNITY

for US Department of Housing and Urban Development

TRITON FOUNDATION, INC.

96 Mt. Auburn Street, Cambridge, Massachusetts 02138

This Urban Planning Research and Demonstration Project was made possible through a contract awarded by the Department of Housing and Urban Development, under the provisions of Section 701 (b) of the Housing Act of 1954, as amended, to the Triton Foundation, Inc

TABLE OF CONTENTS

TRITON CITY : A Prototype Floating Community

Summary

The objectives of Urban Planning Research and Demonstration Project No. Mass. PD-6 were to investigate the broad technical and economic feasibility of developing the water areas of major cities by floating entirely new communities on the water adjacent to the urban core.

Over 80% of metropolitan areas with a population of 1,000,000 or more are situated near bodies of water which are sufficient to accommodate floating cities. Most have a depth of water adequate for shipping (25-30 feet) and relatively sheltered harbors. At these depths, a maximum average height of 20 stories can be floated.

Siting of the city on water gives a unique opportunity for circumventing some of the constraints which currently limit full use by the construction industry of the potential of production technology. With the sea as highway, an entire unit can be built in another location - such as a shipyard or dry dock - and then towed to its site in one piece. By employing a large, existing construction facility of this kind, the economies of shop fabrication can be brought to bear on the construction problems which are traditionally soluble only at the final site location.

Both structurally and organizationally, it is most sensible to provide relatively small (in terms of city sizes), discrete neighborhood platforms - roughly the size, though not necessarily the same proportions, of large liners and tankers. Such platforms will be up to four acres in area and house as many as 5,000 people. In this way, the structural elements are kept to sizes which can be reasonably handled by existing shipyard facilities, and movement of the platforms into place is easily accomplished. Larger town and city complexes can then be made by linking these platforms at the final location.

Maximum structural efficiency will also be effected by considering the platform (which will be of steel or concrete) and its "superstructures" - those portions of the buildings rising above the platform - as a complete framework (in technical terminology, a "megastructure). At the same time that this offers the best structural solution, it also makes possible the most flexible distribution of spatial uses : the requirements for large, open spaces are met, and needs for smaller spaces can be readily accommodated by light weight structures filling in the larger ones. The infilling components (apartments, classrooms, stores, offices) are factory produced as complete, finished units before they are fitted within the frame. This prefabrication of elements is a means of approaching the efficiency of the automobile industry in assembly line production and of achieving similar economies. Additionally, it will be possible to make subsequent changes by removing outmoded units and replacing them with new ones without disturbing the overall disposition of the city.

A whole neighborhood can be treated as a single building functionally and all mechanical services, including water, sewerage and waste, power, and heating and air conditioning, centrally provided. This has two advantages, the first that no duplication of costly central plant equipment for individual units is required, and the second that distribution of services is much more efficient.

The basic unit of Triton City is a neighborhood sized community which will accommodate 3,500 to 6,500 people. This unit, averaging 5,000 residents, is the size required to support an elementary school, a small supermarket, and local convenience stores and services. Three to six of these neighborhoods, with a population of 15,000 to 30,000, will form a town. At this point, an additional platform including a high school, more commercial, recreational and civic facilities, and possibly some light industry, will be added. Given this system of aggregation of separate modules, flexible arrangements of total communities up to 100,000 persons can either grow gradually, starting with a cluster of two or three neighbor hoods, or be built up with great rapidity. When the community has reached the level of three to seven towns (90,000 to 125,000 population), it becomes a full-scale city and incorporates a city center module containing governmental offices, medical facilities, a shopping center, and perhaps some form of city-based activity like a community college or specialized industry.

Should the requirements alter after the city has been established, units can be added or subtracted from the total as change or growth may dictate, and not cause disruption of the entire fabric. Cities can develop the facilities incrementally, without having to make a greater expenditure than is actually warranted at any given point in time.

By designing a megastructure (i.e., an entire framework of structure and services) for high density occupation, great economies in transportation, services and other utilities can be realized. The economy is not only monetary, but also in conservation of open space for recreation and in easy access to the core city by highways and rapid transit. Preliminary cost estimates indicate that the whole fabric (including housing, schools and other community facilities, all services, roads and utilities) can be provided at an expense of $8,000 per person (based on a density of 300 dwelling units per acre).

TRITON CITY: A Prototype Floating Community

Suppose you could choose from all possible alternatives exactly where you wanted to live? Where would that be? In one of the new, luxurious, downtown high-rise apartment buildings? In one of the fashionable suburbs, exurbs or new towns? Or in a renovated town house in a rediscovered center city residential neighborhood?

Many of the fortunate who can afford freedom of choice and who have tried one or two of these possibilities are turning to get another, one which offers the advantages of urban convenience and suburban open space. This is waterfront living in the central city.

Over 80% of our major metropolitan areas -- those with a metropolitan population of more than one million persons -- are situated on or near large bodies of water: the two oceans, the Great Lakes, or continental rivers. This is not surprising since they grew at a time when water-borne shipping and transportaion were essential to the success of new settlements. The potential for waterside development then exists in a variety of American cities: the very large ones like Boston, Chicago, Cleveland, Detroit, Los Angeles, New Orleans, New York, Philadelphia, and any number of smaller cities which border on water.

Today, partly in conjunction with the general concern with urban renewal, but also as a function of heightened interest in water sites, a number of major cities have begun large-scale programs to redevelop their hitherto badly neglected waterfronts. Most notable for this planning are the eastern seaboard cities of Baltimore, Boston, New York and Philadelphia. What has occurred in connection with this trend is a rebirth of interest in land on and near the shoreline. Outmoded, unused wharfs, piers and docks are being reclaimed for use as commercial and recreational sites; warehouses are being converted to luxury apartments. In some cities the most promising area is just that waterfront which a few years ago had been one of the worst problems, and waterside residence is becoming as prestigious as suburban estate housing. Just as the garden suburb was the prevalent notion of the post-war years, the idea of the waterfront community may well be the focal point for the 1970's and 1980's.

But the existing supply of urban waterfront properties is severely limited, the structures very old, and renovation and remodeling extremely expensive; consequently, rents are prohibitively high. As a result, given existing circumstances, people with lower and middle incomes are left with only two choices: stay in the city under unpleasant conditions of dirt, overcrowding and dilapidation or move out and incur a daily round of commuting ranging anywhere from one to three hours.

It is in this context that investigation is being made of the technical and economic feasibility of developing the water areas of major cities by creating entirely new communities on the water adjacent to the urban core. Since most of our metropolitan areas gew up around port facilities, they possess the necessary deep and sheltered water close to shore. In most cases, the available water depth (about 30 feet) is sufficient to float an average load of 20 stories above the water surface. This means that if some sections of the floating structures were lower than 20 stories, others could be still higher.

Findings of this study indicate that it is possible to provide waterfront living for large numbers of city dwellers on floating communities at the shores of our major cities. What does this mean to the average metropolitan area resident? It means that he could have an apartment which is not only close to the heart of the city but also offers the glamorous view, the open vista, the light, sun, and cool, fresh breezes of the waterfront. It means that he could combine the open space views of the suburbs with the conveniences of and rapid access to the downtown area. It means that he could save hours of commuting time.

Certainly the concept is an exciting one. How can it be translated into reality? Actually, the technology necessary to build a floating city is already in existence. We have been building super liners for many years that accommodate populations of the size of entire towns (The S.S. United States, for example, carries 3,000 persons, including crew and passengers). Super tankers are now being constructed which weigh over 200,000 tons dead weight (The 5,000 person neighborhoods will weigh 150,000 tons). Floating platforms for oil derricks and oceanographic experiments have long been in successful operation - some in conditions of unprotected water which do not trouble our city sites. The individual techniques are there and have proven themselves. It is in the combination of these methods that the essence of the technological breakthrough lies.

The basic unit of Triton City is a neighborhood sized community which will accommodate 3,500 to 6,500 people. This unit, averaging 5,000 residents, is the size required to support an elementary school, a small supermarket, and local convenience stores and services. There are two kinds of neighborhood modules designed for the city. One is composed of a string of four to six small platforms, each holding about 1,000 people; the other is a larger, triangular shaped platform which will be of higher density and have capacity for as many as 6,500. Three to six of these neighborhoods, with a population of 15,000 to 30,000, will form a town. At this point, an additional platform including a high school, more commercial, recreational and civic facilities, and possibly some light industry, could be added.

Given this system of aggregation of separate modules, flexible arrangements of total communities up to 100,000 persons can either grow gradually, starting with a cluster of two or three neighborhoods, or be built up with great rapidity. When the community has reached the level of three to seven towns (90,000 to 125,000 population), it becomes a full-scale city and incorporates a city center module containing governmental offices medical facilities, a shopping center, and possibly some form of special city-based activity like a community college or specialized industry.

Should the requirements alter after the city has been established, units can be added to or subtracted from the total as change or growth may dictate, and not cause disruption of the entire fabric. Cities interested in the Triton City adjunct, but unsure of how well it will work for their particular problems, could develop the facilities incrementally, without having to make a greater expenditure than would be actually warranted at any given point in time. Additionally, the modular concept of physical organization would allow each city to develop its own floating configurations in accord with its individual needs and the dimensions of its water basin (e.g. In one case, the floating city might be a line of platforms stretched along the shoreline of the city, while in another, the configuration could be a tight cluster which comes close to the shore at only one or two places).

One of the important attributes of Triton City is that it offers "the best of two worlds": the dynamic quality of life in a milieu of urban high density and the view of immediately adjacent open space which is traditionally the province of suburban and rural areas. There would also be playfields and parks fitted in between neighborhood modules. And, since the community is intended as a city complement, it would have all of the existing urban amenities, including entertainment and educational and cultural activities, to draw upon as well.

The prototype density used for the Triton City communities is 300 dwelling units per acre. This high density of population would economically support some form of transportation between the community and the city core. While there is automobile access to and from the floating platforms and parking for residents' cars, it is anticipated that movement from platform to platform (i.e., from one neighborhood to another) would be a walkable distance or accomplished by public transit. In order to discourage automobile congestion in the floating city, there is little provision for transient parking. All wheeled vehicles are restricted to a single level in the city complex, which is segregated from pedestrian areas. At this level are truck loading and unloading facilities, transit stations, and ramp access to parking garages It is probable that the transit system would be rubber tired and would circulate on the same roadbed as automobiles, buses and trucks. However, possible use of a system such as the Westinghouse sky bus being considered.

Because the megastructure constitutes a neighborhood entity, some new departures in aesthetics and safety can be realized. All parking is within the flotation, so that one major contemporary eyesore, the parking lot, is removed from view. Since wheeled vehicles are not permitted above the entrance level, the streets would be safe for pedestrians. Every neighborhood child could walk to school- and in no danger of being run over. As another precaution, the elevators and stairs, which are housed in vertical towers, would have glazed sides, so that everyone inside is visible at all times. The installation of vertical circulation facilities in three centrally located towers also means that they could be surveyed from one vantage point and that they would be intensively used, thus dually insuring the safety of residents. Moreover, there are no dangerous alleyways and no hidden access to any dwelling, as all doors are directly on the streets, which are wide, straight and easily patrolled.

The dwelling units all have frontage directly on the water, and the exterior faces slope slightly backward so that apartments on the higher levels look on the garden terraces of those below rather than, clifflike, straight down to the water. Apartments on the upper levels give magnificent views while those on the lower levels offer an equally fascinating closeness to the water.

The front doors of the dwelling units open onto broad (about 18 feet wide) "streets in the air" that are solely for pedestrian use and very much resemble the promenade decks of ocean liners. These streets are connected by bridges to the schools, shops and other community facilities, which are in the interior portion of the megastructure. At the higher levels, the apartment units surround and enclose the village square, a public space open to the sky. The many roof levels of the structure are terraced and landscaped for various kinds of recreation. Some contain tennis courts; others provide nursery or play areas; still others are small parks for walking, sitting or reading.

This arrangement of residential, institutional, commercial and recreational facilities creates a new kind of townscape. In contrast to the ordinary concept of the town as horizontally developed, Triton City operates on the basis of both horizontal and vertical correspondences and accesses. High-rise apartments and office buildings demonstrate the principle of vertical organization; but there, of course, only one or two uses are involved, and all others remain on the horizontal plane. The Triton City megastructure organizes residential and commercial space vertically, but the commonly ground-oriented community facilities and services are available on all levels.

This organization of city functions allows for other human amenities beyond the operational requirements. Since there is no separation, either horizontally or vertically, of diverse facilities, a great many different kinds of spaces can be found within the megastructure. It is this constantly shifting pattern of use and the concomitant interplay of spaces which makes the most successful parts of our existing cities so exciting, for not only is the diversity interesting, but is also keeps the scale of activity appropriate to the human being. Thus, the community areas of Triton City, the interiors which are designed for human passage and leisure would be engaging and would tend to encourage people to pass through, to stop and linger, and to participate in the activities.

Because the environment is inviting spatially, functionally, and in terms of safety, people could experience the pleasure of walking in urban spaces which is now limited to so few areas, such as Beacon Hill in Boston. Nor need this enjoyment be restricted to the inside of the community. Since they would be so close to the mainland, residents and visitors could walk to and from their floating city. This is one of the most attractive aspects of urban living to many of the people who choose to reside in the hearts of our major cities. And to be able to walk not only to and from work, but also to beaches, boating and other recreation increases the desirability of the floating city by yet another order of magnitude.

The siting of the city on water gives a unique opportunity for circumventing some of the constraints which currently limit full use by the construction industry of the potential of production technology. With the sea as highway, an entire neighborhood unit can be built in another location - such as a shipyard or dry dock - and then towed to its site in one piece. By employing a large, existing construction facility of this kind, the economies of shop fabrication can be brought to bear on the construction problems which have been traditionally soluble only at the final site. location.

Both structurally and organizationally, it is most sensible to provide relatively small (in terms of city sizes), discrete platforms - roughly the size, though not necessarily the same proportions, of large liners and tankers. Such platforms would be up to four acres in area and house as many as 5,000 people. In this way, the structural elements are kept to sizes which can be reasonably handled by existing shipyard facilities, and movement of the platforms into place is easily accomplished. Larger town and city complexes could than be made by linking these platforms at the final location.

Maximum structural efficiency is effected by considering the platform and its "superstructures" - those portions of the buildings rising above the platform - as a complete framework (or in technical terminology a "megastructure"). At the same time that this offers the best structural solution, it also makes possible the most flexible distribution of spatial uses: the requirements for large, open spaces are met , and needs for smaller spaces can be readily accommodated by light weight structures filling in the larger ones. The infilling components (apartments, classrooms, stores, offices)would be factory produced as complete, finished units before they are fitted within the frame. This prefabrication of elements is a means of approaching the efficiency of the automobile industry in assembly line production and of achieving similar economies. Additionally, it would be possible to make subsequent changes by removing outmoded units and replacing them with new ones without disturbing the overall disposition of the city.

A whole neighborhood can be treated as a single building functionally and mechanical services, including water, sewerage and waste, power, and heating and air conditioning, centrally provided. This has two advantages, the first that no duplication of costly central plant equipment for individual units is required, and the second that distribution of services is much more efficient. Use of the interior zone of the megastructure allows for a much more closely controlled environment than would be feasible in a usual sort of development - for temperature and humidity as well as for access and circulation. Moreover, the economies of centralized services mean that air conditioning of private wellings; automatic trash and garbage disposal; and adequate lighting of corridors, streets and other public areas are simple inclusions. (These are frequently not found, even in "luxury" apartments and developments, except as costly extras!)

Steel or concrete for building floating plat forms is presently available and has been adequately tested to assure economy and safety. There are examples of concrete boats which have been in service for over 40 years with no signs of weakening, leakage or undue corrosion or fouling to the hull. In addition to ordinary and special steel alloys for the hulls, newer protective coatings for steel have proven durable in water applications.

Just as the depth of draft and "attitude" in the water are controlled in ships and submarines, floating plat forms can be stabilized by pumping water or air through the flotation to compensate for changes in loading and for wind and wave effects. The platform can be made stable enough so that no more movement would be experienced than on an ocean liner in harbor.

Accurate cost comparisons of floating communities to orthodox developments on land are difficult to make because it is almost impossible to obtain any single complete, consistent and accurate set of cost figures for existing communites and developments. Cost analyses for such developments are never done on the basis of total implementation, so that available figures are usually incomplete, inconclusive and sometimes misleading. However, initial estimates indicate that at the prototype density, the per person costs for the floating communities would be competitive with the per person expense for conventional developments (more so when the extra amenities of the floating platform solution are added to the costs of usual suburban type developments) and comparable urban renewal projects in large cities.

But perhaps the most significant advantages of Triton City lie within the social context of the possibilities it can offer. Because of the economies inherent in the planning and construction of the community, the costs can be low enough for many more people to have the chance to live on the water. For the first time in our society, the pleasures of waterside living would be open not just to the rich or "privileged" classes. Moreover, the standards of living would be higher than they are in almost all other kinds of residential development, with generous spatial allotments, convenient access, and clean, fresh, conditioned air circulated throughout all the living spaces. And importantly, the public areas, with adequate lighting and intensive use, would be far safer than our downtown streets at night (crime rates having been shown to be related directly to darkness and the absence of people on the street). Since it is so close to the existing city and can economically support a public transportation system, the community could enjoy both the benefits of urban living and access (as little as 5 or 10 minutes to the downtown area) and the pleasant qualities of a seaside vacation home. There is even the possibility for marinas to be included in each neighborhood, both as private docks and as public marinas for rental of boats to visitors or permanent residents.

We are now confronting the problem of expansion in our metropolitan areas at a scale vaster than ever previously contemplated. At the same time, we are in need of renewing or rehabilitating substantial portions of the existing city fabric, posing the additional complication of what to do with the people, services and facilities housed there. This problem of "relocation" has proved to be a very costly consideration in renewal processes and has, indeed, been one of the reasons why urban renewal has become both so expensive and so piecemeal.

From time to time, densely settled cities do "discover" sizable construction sites by reconverting military land to civilian use, by reuse of rail yards, by filling shore areas, or more frequently, by assembling parcels devoted to "lower uses" to some higher use. This last technique, often used in connection with urban renewal, has been attacked as a form of "Negro removal" or an assault on the inadequate supply of low income housing units available to the poor. One of the advantages of Triton City is that it entails no disruption of existing neighborhood patterns or loss of dwellings. It will add to rather than diminish the supply of housing and augment the number of cohesive neighborhoods in and near the urban core.

While there is no single answer to complicated urban questions, the Triton City concept can at least offer another choice of environment to live in and help to remove some of the present stumbling blocks. The floating community can be envisaged as a means of accommodating the multiplying metropolitan populations which are threatening and choking our existing cities. It can equally well be used to house persons - either temporarily or permanently - from proposed renewal areas: by eliminating, in this way, some of the difficulties and delays of relocations, subsequent urban renewal projects might be done a bit more cheaply. Also, the flexible nature of both the arrangement of modules and their internal organization is such that,within the "megastructure" context, adjustments can be made and the floating community brought "up to date" in response to changes or developments within the city.

The idea of living in a spacious, clean apartment right on the water, yet immediately convenient to shops and schools, within minutes of downtown and at reasonable cost, is unquestionably an attractive one. There are a great many people who could be interested in this kind of opportunity for its own sake. Others, who ordinarily would wish to live in the suburbs for proximity to recreation and open space, could enjoy these and the added benefit of being close to place of employment, with a resultant increase in leisure time brought about by elimination of commutation between suburb and city. Those who already live in the city, and certainly those who govern it, will welcome the introduction of new, dynamic, effective means of urban revitalization.

10. LOCATIONAL POSSIBILITIES AND SITE ENVIRONMENT

11. Locational Possibilities

The floating cities have been designed for location off-shore from high density urban centers, preferably those which are older and in need of renewal. In these circumstances, the communities could function dually relieving pressures of rising population and inadequate housing, to help expedite the urban renewal process.

According to the U. S. Census of 1960[a], there are 28 metropolitan areas in the United States having more than one million inhabitants. Of these, 23 are located on some body of water (See list below). Some (about 7, as shown below) are cities where the depth is too shallow or where the width is too narrow to accommodate a floating development without impairing either navigation or the natural shoreline. The remaining 16 are all reasonably situated in terms of size and depth of water and adequate approaches.

U.S. Metropolitan Areas[a]	No Water[b]	Insufficient Water[b]	Suitable Location[b]
Atlanta, Georgia	X		
Annaheim/Santa Ana/ Garden Grove, California	X		
Baltimore, Maryland			X
Boston, Massachusetts			X
Buffalo, New York			X
Chicago, Illinois			X
Cincinnati, Ohio		X	
Cleveland, Ohio			X
Dallas, Texas	X		
Denver, Colorado	X		
Detroit, Michigan			X
Houston/Galveston, Texas			X
Kansas City, Kansas		X	
Los Angeles, California			X
Miami, Florida			X
Milwaukee, Wisconsin			X
Minneapolis/St. Paul, Minnesota		X	
Newark, New Jersey		X	
New Orleans, Louisiana			X
New York City, New York			X
Patterson/Clifton/Passaic New Jersey	X		
Philadelphia, Pennsylvania			X
Pittsburgh, Pennsylvania		X	
San Diego, California			X
San Francisco, California			X
Seattle, Washington			X
St. Louis, Missouri		X	
Washington, D.C.		X	

12. Site Considerations

For preliminary design purposes, the structure is assumed to be located in a protected body of water, close to the shore of an existing city. A water depth of 30 feet below mean low water is available in most such locations.[b] Thus, a maximum draft of 30 feet in calm water is imposed for the floating support structure. The 16 cities mentioned in paragraph 11. all possess facilities for major shipping, which requires a channel draft of 30 feet. Given this depth of water, the floating megastructure could attain an average height of 20 floors, or 200 feet at 10 feet floor to floor, as shown below:

Average weight:	40-50 psf dead load
Residential floor:	50-40 psf live load
	90 psf total load per floor

Maximum height: 30' water X 62.4#/cu.ft. = approximately 1800 psf

1800 psf / 90 psf per floor = 20 floors residential load

20 floors x 10' /floor = 200' height

13. Site Environment

The most severe wind and sea conditions that may be experienced at a particular site are the criteria which determine anchor requirements, freeboard, tidal elevation changes at land connection, pitch and roll effects, and hull bending due to waves. These conditions will vary considerably from one site to another. However, the following range of values is useful for initial discussion and preliminary design:

(1)	Wind -- velocity pressure 20 to 35 psf (use 30 psf for preliminary design)	
(2)	Tides -- normal range (astronomical)	2 ft. to 10 ft.
	Storm tides -- additive in hurricane (Use 18 ft. max. tide range for preliminary design)	8 ft. to 12 ft.
	Tidal velocity of flow (Use 2 ft./sec. for preliminary design)	1 to 2 ft./sec.
(3)	Wind waves with 35 ft. bottom depth, 100 mph maximum long-duration wind velocity --	
	For large body of water -- significant height*	12 ft.
	significant period**	8.5 sec.
	For body of limited extent (one mile fetch) -- significant height	6 ft.
	(Use 8 ft. height and 8.5 sec. period for preliminary design)	
	Maximum velocity of flow in long waves (Use 3 ft./sec. for preliminary design)	2 to 3 ft./sec.
(4)	Temperature range Ice may form in some locations exposed to low temperatures and still water.	-20°F to 100°F

Meteorological records and tide tables for an individual site should be used for final design of any specific project. Wave heights and period may be calculated using meteorological data for the site and charts given in d. (See also paragraph 51.)

---

* Significant height is the average height of the highest third of all the waves present in the wave train and is considered approximately equal to the height usually reported in observations. [c]

** Significant wave period represents mean period of significant wave height.[c]

14. Vertical Clearances

Since the structures will be fabricated elsewhere and floated to their destinations, the question of vertical bridge clearances becomes a consideration. Many major harbor bridges have vertical clearances of over 200 feet, e.g. the George Washington, 211 feet; the Verrazano Narrows, 217 feet; the Golden Gate, 232 feet.[b] However, the Coast Guard regulations require that maximum height for any floating vessel be limited to 160 feet above water level.[e] When the Triton City structure is to be higher than 16 stories, the top stories will have to be added in situ. This does not pose any serious problem either to fabrication or construction. Moreover, there will be many cases (perhaps most) where the 160 feet height limit is not exceeded.

10. References:

a ______________, Pocket Data Book, USA. 1967. U.S. Bureau of Census, 1967.

b ______________, various maps, U.S. Coast & Geodetic Survey.

c A. T. Ippen, Editor, Estuary and Coastline Hydrodynamics. McGraw-Hill, 1966.

d ______________, Shore Protection - Planning and Design. U.S. Corps of Engineers, Coastal Engineering Research Center, Technical Report No. 4, 3rd Edition.

e ______________, Rules and Regulations for Passenger Vessels. U.S. Coast Guard., C.G. 256.

20. POPULATION STATISTICS AND ORGANIZATION

21. Developing a Balanced Population

The patterns of racial and income segregation found in central cities tend to be greatly accentuated in suburbs. The massive post World War II population shift, which combined a heavy growth in suburban communities and an exodus of middle income Caucasians from the urban core, has tended to increase the physical distance between races and classes and diminish the interaction and intermingling in the public schools and in other areas. There is considerable sentiment against this type of segregation and a developing body of law aimed at eliminating racial imbalance and preventing discriminatory practices in suburban zoning. On the books but still not fully funded are such programs as rent supplements, and there is a proposed negative income tax to provide the poor with the financial wherewithal to purchase better housing. The off-shore cities can be designed to embrace the spectrum of social and economic classes and races through rent supplements or other types of financial aid for poorer families.

There are, however, certain limits to the diversity of population desirable in an experimental high density settlement. It must be reiterated that the new communities cannot and should not be expected, at least in their initial stages, to assume the responsibility of dealing with multi-problem families or the mentally or socially unstable. A policy of exclusion based on demonstrably anti-social behavior may well be a prerequisite for success in achieving the objectives of the new communities.

22. Population Breakdowns: Reference [a]

	Families/ Household: %	Persons/ Household	Persons/ Family	Unrelated Individuals / 100 Families
U. S. Average	85%	3.29	3.65	29.8
Mass.	84	3.23	3.61	33.1
Conn.	87	3.27	3.57	27.6
New York	82 1/2	3.2	3.49	33.5
Illinois	84	3.18	3.55	30.6
Michigan	87	3.42	3.73	25.4
Wash., D.C.	69	2.87	3.52	80.7
California	80	3.05	3.50	39.9
TRITON estimates for urban average	80%	3.1 (Ref. b)	3.4	50

23. Family Size

	U.S. Average (Ref. c) % of family population	TRITON estimates for urban populations % of family population
2 persons per family	32.2%	37.6%
3 persons per family	20.7	24.8
4 persons per family	19.9	18.2
5 persons per family	13.2	10.0
6 persons per family	7.0	4.7
7 & more persons per family	7.0	4.7

Median age: 28.3 years (Ref. d)

24. Age Groups

(TRITON assumption is that U.S. average is accurate enough since 90% of population is urban.)

	U.S. Average (Ref. e)
Under 1 year	2.1%
1- 4 years	8.7
5-13 years	18.4
14-17 years	7.4
18-21 years	5.8
22-64 years	49.7
65 and over	9.3

25. Organization of City Facilities

Organization of city facilities can be divided functionally into the following basic categories:

> 3,500-6,500 population: minimum population required to support elementary school and branch supermarket = VILLAGE MODULE[f]
>
> 15-30,000 population: next higher level of population required to support high school, local government facilities, some specialty commerce and service/support activities = TOWN MODULE [g]
>
> 90-120,000 population: approximate size required to support shopping center - also requires additional community and commercial support activities. Requisite size for specialized local activities and facilities = CITY MODULE [g]

When this facility organization is carried into physical form, the following patterns emerge:

> 1 village module: 3,500-6,500 population = Neighborhood
>
> 3-6 village modules: 15-30,000 population = Town
> requires inclusion of additional service/support facilities in separate module
>
> 2-6 town modules: 90-120,000 population = City
> requires another higher level addition of service/ support facilities in separate module

Organizing the city in this manner has several distinct advantages:

1. Implementation of the city can be incremental.
2. Flexibility of city configurations is possible to accommodate different capacities and site requirements.
3. Specialized local requirements can be fulfilled by the inclusion of specialized modules.

4. Initial trials can be undertaken on a minimum involvement basis, and feedback experience can then be applied to adjustment of use of further modules.

26. A Note on Population Densities

The proposed population density of about 930 persons (or 300 dwelling units) to the acre in Triton City is clearly different from the 1,000 persons (or 270 dwelling units) to the square mile found in some outlying exurban townships. As an urban density it is also considerably higher than the maximum recommended density of 95 dwelling units[h] or about 350 persons per acre given by the American Public Health Association. However, Manhattan residents in new 13-story buildings typically live in densities ranging from 300 to 500 dwelling units per acre, far in excess of Triton City.[h]

Perhaps more to the point in density comparisons is Jane Jacobs' assertion, based on empirical observation, that a minimum of 200 dwelling units to the acre is required for a healthy, energized urban neighborhood. She points out that Brooklyn Heights, the most generally admired part of that borough, has the highest density in Brooklyn.[i] In Boston's North End, the density of 275 units to the acre is approximately equal to the proposed off-shore cities. In contrast, Jacobs describes slum area densities as typically ranging downwards from 100 units to only 15 units to the acre.[h]

27. Facility Requirements: [k]

	Per 1,000 Population
Retail markets	8.7
Wholesale markets	1.6
Services	5.5
Telephones	478
Pieces of mail	3.6×10^5
Electrical consumption: residential	1.4×10^6 KWHR
all other	3.5×10^6 KWHR
Transit	4.3×10^4 person trips/day
Water consumption[l]	$5\text{-}10 \times 10^4$ gal/day

	Per 100,000 Population
General hospital admissions	145
Total days in hospital	1331
Doctors	151
Dentists	56
Nurses	306

20. References:

a ______________, Statistical Abstract of the United States, 1965. U.S. Bureau of Census, Washington, D.C., 86th ed., 1965, Table 39.

b ibid., Table 475.

c ibid., Table 38.

d ibid., Table 19.

e ibid., Tables 18 and 19.

f ______________, Planning the Neighborhood. U.S. Public Administration Service, 1960.

g Clark Applebaum, Store Location and Development Studies. Clark University, 1961.

h Bernard J. Frieden, "Local Preferences in the Urban Housing Market" Journal of the American Institute of Planners, November, 1961 p. 332.

i Jane Jacobs, The Death and Life of Great American Cities. Random House, New York, 1961.

j ______________, Statistical Abstract of the United States, 1965, Table 476.

k ______________, Pocket Data Book 1966 . U.S. Bureau of Census, Washington, D.C., 1967.

l Ramsey and Sleeper, Architectural Graphic Standards. John Wiley and Sons, New York, 5th ed., 1962.

30. AREA AND SPACE REQUIREMENTS

31. Space Allocations

1) Residential:

Ref. [a]:

FHA minimum standards: approximately 120 sq.ft./person net livable floor area

FHA minimum standards: storage space approximately 40 sq.ft./person

total: 160 useable sq.ft./person

TRITON estimates

Provide 200 sq.ft./person net useable floor area

Add 50% for structura, partitions and exterior circulation

= 100 sq.ft./person

Total: 300 sq.ft./person (gross area)

2) Schools:

Ref. [b]: Total sq.ft./student: includes classrooms, auxiliary, service

High	116
Low	75

Ref. [c]: Recommended minimum size of school buildings and playground: 5 acres

TRITON estimates

	Square feet/student
Elementary school	80
Junior High school	100
High school	125
Community college	200 (includes dormitory space)

3) Hospitals:

Ref. [d]:

2 1/2 beds per 1,000 population for population less than 25,000
4 beds per 1,000 population for population greater than 25,000

Ref. [e]:

average 750 sq.ft. per bed: gross area
average $30 per sq.ft.: cost

TRITON estimates

5,000 pop.:	emergency medical facilities only
20,000 pop.:	small clinic for medical offices and support facilities
100,000 pop.:	complete medical clinic facilities, including emergency hospital provisions

N. B. : major hospital facilities assumed to be in adjacent center city core; hospital capacity to be expanded there to take advantage of large, specialized facilities.

4) Churches

Ref. [f]:

approximately 0.7 to 0.93 per 1,000 population
(487 in Worcester county, Massachusetts: population 597,470)

Ref. [g]:

30 churches/75,000 population: Reston = 1 church/2,500 population

TRITON estimates

1 small church for 2,000-2,500 population located in neighborhood, with additional and larger churches located in town and city centers

5) Commercial

Ref. [h]: Shopping Center for 5,000 population:

Ground area of buildings	25,000 sq.ft.
Customer parking	50,000 sq.ft.
Service station	24,000 sq.ft.
Circulation, services & setbacks @ 25%	25,000 sq.ft.
	124,000 sq.ft.

Ref. [g]:

Reston: 1,000,000 sq.ft. for 75,000 population
13.3 sq.ft./person

Ref. [i]:

1,000,000 sq.ft. for 30,000 population
30 sq.ft./person

TRITON estimates

5,000 pop.:	local and convenience shopping only: food, drugs, services 5 sq.ft./person
20,000 pop.:	additional specialty shops and branch chain stores 7 sq.ft./person - stores 1 sq.ft./person - service station
100,000 pop.:	department, furniture, etc. stores: level of secondary shopping center 8 sq.ft./person
TOTAL:	21 sq.ft./person

6) Multi-family Development Neighborhood (Conventional): Ref. [h]:

population 5,000; 1,375 families

School	2.2 acres
Playground	6.0 acres
Park	6.0 acres
Shopping center	3.0 acres
Community facilities	1.9 acres
	19.1 acres

7) Access: Ref. [h]

	Walk	Transit
Jr. high school	3/4 - 1 mile	15-25 min. - free
Sr. high school	1-1 1/2 mile	20-30 min. - 10¢ max.
District center - incl. health	1-1 1/2 mile	20-30 min. - 10¢ max.
Employment		20-30 min.
Urban center		30-45 min.
Outside recreation		30-60 min.
Athletic playfields	1-1 1/2 mile	

8) Open Space

Ref. [j]:

British new towns	15-30%
Reston	32%
Columbia	27%

Ref. [k]:

for 100,000 population, require 1 acre/100 population
for less than 10,000 population, require more than 1 acre/100 population

Ref. [j]:

for dense urban population, 50,000 sq.ft./1,000 population
is acceptable

32. Area Requirements for Village Module "A"[1]: 4800 population

Facility	# of persons	sq. ft./person	# of units	sq.ft./unit	total sq. ft.	% of total
RESIDENTIAL:	4800	300	1480	930 avg.	1,440,000	60
Unrelated persons + :	520	300	250			
25% single *	120		120	426		
25% double *	80		40	600		
50% 3 & 4	320		90	900-1200		
Families:	4280	300	1230			
2 persons*	884		442	600		
3 persons	985		305	800		
4 persons	936		234	1000		
5 persons	695		129	1000		
6 persons	340		60	1300		
7 & more	440		60	13-1800	107,200	4.5
INSTITUTIONAL:	4800	21.7				
School: grades 1-9	920 students	80	1		73,600	
Library:		included in schools	2nd town center			
Museum:		at city	center only			
Hospital: medical		3	1		14,400	
Church:		2	2-3		9,600	
Clubs:		2	2		9,600	
Auditoria:		included in	schools		9,600	0.4
GOVERNMENTAL:	4800	2				
Fire:		in town center				
Police:		in town center				
Post Office:		in town center	1		9,600	
Administration:		2			28,800	1.2
COMMERCIAL:	4800	6			24,000	
Retail: sales & services		5				
Wholesale:		at city center only			4,800	
Offices: professional		1				
INDUSTRIAL:		in town center			240,000	10
RECREATION/OPEN SPACE:	4800	50				
Parks, playgrounds, public space					74,000	3.1
PARKING:						
@ 1/4 x # d.u.			370	200	74,000	
SERVICES:		5			24,000	1.0
Mechanical, electrical & plumbing ducts & chases						
SUB-TOTAL:	4800				1,923,600	
CIRCULATION @ 25%		101			480,900	20.0
TOTAL.		565.7			2,404,500	100.2

VOLUME: Residential @ 10' FL/FL - 14,400,000
Non-residential @ 20' FL/FL = 19,290,000
TOTAL - 33,690,000

\+ Distribution of unrelated persons has been arbitrarily subdivided to give correct average number of persons per d.u.

* 40 single and 80 double unrelated persons and 10 2 person families shifted to town center module

1 NOTE: Other villages are composed of rectangular platform modules of 800 to 1,000 population each. 4 to 5 of these modules constitute a village with the same proportionate allocations of space usage as for module "A".

Area Requirements for City Center Module

Facility	# of persons	sq.ft./person	# of units	sq. ft./unit	total sq. ft.	% of total
RESIDENTIAL:	2000	300	970	620 avg.	600,000	34
Unrelated persons + :	723	300	463			
25% single +	231		231			
25% double *	430		215			
50% 3 & 4	62		17			
Families:	1277		507			
2 persons *	730		365			
3 persons	219		73			
4 persons	160		40			
5 persons	90		18			
6 persons	42		7			
7 & more	36		4			
INSTITUTIONAL:	20,000	11-1/4			225,000	12.8
High school	1480 students	125	1		185,000	
Library: adjunct to school		1/4	1	5,000	5,000	
Museum: at city center						
Hospital: clinic		1	1	20,000	20,000	
Church		1/4	1-2	2-5,000	5,000	
Clubs		1/2	2-4	2-5,000	10,000	
Auditorium:		included in	high school			
GOVERNMENTAL:	20,000	1-1/4			25,000	1.4
Fire: station		1/4	1	5,000	5,000	
Police: precinct		1/4	1	5,000	5,000	
Post Office: branch		1/4	1	5,000	5,000	
Administration:		1/2	1	10,000	10,000	
COMMERCIAL:	20,000	8-1/2			170,000	9.6
Retail: sales & services		5			100,000	
Wholesale:		at city center				
Office space:		2 1/2			50,000	
Service station:		1			20,000	
INDUSTRIAL:		6			120,000	6.8
R & D and related industrial						
RECREATION & OPEN SPACE:		5			100,000	5.7
PARKING & TRANSIT:		7			150,000	8.5
@ 2/3 x # of d.u.		6.5	650	200	130,000	
Transit station		.5	1		20,000	
SERVICES:		5			40,000	2.3
Mechanical, electrical & plumbing ducts & chases						
SUB-TOTAL:					1,430,000	
CIRCULATION @ 25%		16.5			332,500	19
TOTAL:					1,762,500	100.1

VOLUME: Residential @ 10' FL/FL = 6,000,000 cu. ft.
Non-residential @ 20' FL/FL = 25,250,000 cu. ft.
TOTAL: 31,250,000 cu. ft.

+ Distribution of unrelated persons has been arbitrarily subdivided to give correct average number of persons per d.u.

* 40 single and 80 double unrelated persons and 40 2 person families shifted from village module

34. Area Requirements for City Center Adjunct Module: 100,000 population

Facility	# of persons	sq.ft./person	# of units	sq.ft./unit	total sq. ft.	% of total
RESIDENTIAL:	2,000	500	650	1550 avg.	1,000,000	22
INSTITUTIONAL:	100,000	6-1/4			625,000	13.7
School: college	2,600	200			400,000	
Library: public		1/2	1		50,000	
Museum: marine		1/2	1		50,000	
Hospital: major clinic		1/2	1		50,000	
Church		1/4	2-3		25,000	
Clubs		1/2	5		50,000	
Auditorium		included in	college			
GOVERNMENTAL:	100,000	1.7			170,000	3.7
Fire: precinct		.1	1		10,000	
Police: includes jail & courts		1/2	1		50,000	
Post Office: district branch		.1	1		10,000	
Administration		1.	1		100,000	
COMMERCIAL:	100,000	8.5			850,000	18.7
Retail: sales & services		5			500,000	
Wholesale: local distribution		1			100,000	
Office and professional		2 1/2			250,000	
INDUSTRIAL:		optional -	included in separate module			
RECREATION & OPEN SPACE:	100,000	3			300,000	6.6
includes college & public space						
PARKING & TRANSIT:	100,000	2			200,000	4.4
@ 1 x # of d.u.		1.3	650	200	1 [illegible]0,000	
		.5	1 or 2		70,000	
SERVICES: Mechanical, electrical & plumbing ducts & chases		.5			500,000	11.
SUB-TOTAL:					3,645,000	
CIRCULATION @ 25%:		9.1			911,200	20.
TOTAL:					4,556,200	100.1

VOLUME:

Residential @ 10' FL/FL = 10,000,000 cu.ft.
Non-residential @ 20' FL/FL = 71,124,000 cu.ft.
TOTAL = 81,124,000 cu.ft.

35. Area and Space Allocations Summary

FACILITY	VILLAGE (Module "A") pop. 4,800	TOWN CENTER[1] pop. 2,000	CITY CENTER[2] pop. 2,000	TOTAL SQ.FT. PERSON	TOTAL[3] SQ.FT. x 1,000	% OF TOTAL
	GROSS AREAS IN SQ.FT. x 1,000					
RESIDENTIAL	1380	600.	1000.	316	31,600	53
INSTITUTIONS	92.8	225.	625.	36.1	3606	6.5
School: Grade 1-6	55.1	--	--			
Grade 7-9	22.5	--	--			
Grade 10-12	--	185.	--			
College	--	--	400.			
Other: medical, church, clubs, etc.	19.2	40.	225.			
GOVERNMENTAL fire, police, admin.	9.6	25.	170.	4.9	487	.8
COMMERCIAL AND PROFESSIONAL OFFICES	43.2	170.	850.	25.6	2564	4.3
INDUSTRIAL: R&D	--	120.	--	6.	600	1.0
RECREATION & OPEN SPACE	240.	100.	300.	56.	5600	9.4
PARKING & TRANSIT	74.	150.	200.	24.3	2430	4.1
SERVICE & UTILITIES	24.	40.	50.	17.3	730	1.2
Subtotal	1863.6	1430.	3195.	--	--	--
CIRCULATION @ 25%: vehicle & pedestrian	464.	357.5	799.	118.6	11,866	20.
TOTALS:	2327.6	1787.5	3994.	594.8	59,483	100.3

1. Includes only Town Center - add 4 villages for totals
2. Includes only City Center - add 5 towns for totals
3. Totals include: 20 Villages + 5 Town centers + 1 City center

36. Flexibility

From time to time urbanologists have proposed extensive changes in the city fabric. The creation of government centers, renewed central business districts (with and without malls) and sizeable private office or residential areas represent alterations in the physical urban structure enormous in cost, slow of accomplishment and, once under way, virtually irreversible. Important attributes of Triton City are its locational flexibility; its flexibility of size and rate of growth, which is a function of the modular based systems approach to construction; and, quite possibly, its susceptibility to a wide range of uses, rent levels or ownership arrangements. These can be provided without the necessity for mammoth investment in time, effort and money and without the permanent, or at least long-range commitment implied by most substantial in-town developments.

If the floating city is properly viewed as an experimental operation, it offers the exciting opportunity for ex post facto adjustment to changing circumstances. It can be shifted from one neighborhood to another, or from one shore city to another or, indeed, from one mix of uses to another. Size and rate of growth can be responses to actual current conditions rather than predetermined and possibly erroneous fixed patterns. Various weights can be given to such factors as siting near underutilized public facilities and determination of location and uses to achieve a variety of goals. These may include requirements for office space, for specialized industry or for campus sites. The city can be used for specialized recreation; for expositions and fairs; for providing sites for medical complexes; for temporary or permanent housing to cope with the impact of urban renewal or highway construction; or simply to permit a thinning out of densely settled areas.

There can be little doubt that each floating city will develop distinctive patterns of usage and location in accordance with municipal objectives. One might suggest only one common thread, namely the conversion by municipalities of their presently sealed off, neglected and obsolescent waterfront areas to meet modern community needs.

30. References :

a ______________, Minimum Property Standards for Multifamily Housing. Federal Housing Administration, Washington, D.C., November 1963.

b ______________, The Cost of a Schoolhouse. Educational Facilities Laboratory, Ford Foundation, 1960.

c J.L. Taylor, School Sites: Selection, Development and Utilization. Office of Education, Department of Health, Education and Welfare, 1962.

d Telephone conversation with Dr. Rubenstein, Bureau of Hospital Facilities, Massachusetts Department of Public Health, 7 November, 1967.

e ______________, Hospital Profiles. U.S. Department of Health, Education and Welfare, Public Health Service, 1964.

f Telephone conversation with Dr. Kopper, Massachusetts Council of Churches, 7 November 1967.

g ______________, RESTON Master Plan Report, Fairfax County, Virginia, 1962.

h ______________, Planning the Neighborhood. U.S. Public Administration Service, 1960.

i ______________, Analysis of the Hypothetical Development of a Model New Community. Housing and Home Finance Agency, Washington, D.C., unpublished draft, January 1965.

j Dave Brodeur, Metropolitan Open Space Development. Department of Housing and Urban Development; TRITON meeting with Lenore Siegelman, Department of Housing and Urban Development, 21 November 1967.

k ______________, brochure, Municipal Recreation Administration.

40. TRANSPORTATION

41. Trip Generation Volumes.

In order to discuss a transportation system for Triton City, it is necessary to estimate the order of magnitude of trip interchange which may be expected. Trip characteristics are somewhat simpler than normal patterns since, in this case, little externally-oriented through travel will exist. Basically, there will be two types of trips:

1) Local trips with both origin and destination within the floating complex.

2) Internal to external trips with either origin or destination inside Triton City.

Since the internal employment opportunities will be limited by relatively small amounts of commercial and industrial usage, the major travel movements will be work commuter trips to the core city of the metropolitan area. The short distance projected between Triton City and the adjacent core suggests that work travel will exhibit pronounced peaking characteristics above normal city averages. Work trips generated by each village of 5,000 population will be about 1,750 per weekday. Assuming that 60% of these off-site commuter trips will be made in the morning peak hour, 1,000 trips will result for each village module "A". In terms of a fully developed floating city of 100,000 people, this means a total morning peak hour volume of 18,000 person trips (1,000 from each "A" village and 500 from each town center).

Local movements between village and town centers and between village and city center have also been estimated, based upon known trip generation relationships, particularly for the commercially developed uses; and where applicable, floor area ratios have been used. These computations result in a total local movement between the villages and town centers of the floating community of 12,500 person trips per day. This is approximately 60% of the core area trips which may be expected from a typical urbanized area of 100,000 people.[a] Since the 12,500 trips to the Triton City center are basically nonwork trips, it would appear that there is relatively good agreement between the two sets of figures.

Trip movements between a village module and the town center will be approximately 925 person trips per day. Movement among individual villages has not been estimated since it will depend to a large extent upon the organization of the floating city, the distances between modules, and the efficiency of the transportation system. In any event, this movement will be comparatively small and will not control transportation system design.

42. Alternate Transportation Systems.

Three types of transportation systems are possible for circulation within the floating city and between it and the metropolitan core. These are:

1) A conventional automobile-highway system.
2) A public transit system.
3) A combination automobile and transit system.

The proposal for an automobile dominant transportation system arises from a desire to utilize two levels of subsurface platform volume for parking automobiles. Assuming a conservative vehicle ownership rate, we can expect approximately 30,000 automobiles operated by the residents of the floating city. It appears that a limited highway network linking the residential modules, town centers, and city center will be required for freight movement, and this further justifies the consideration of some use of the passenger car for circulation.

The average city of 100,000 people has a highway network of at least 200 miles. Construction of a network of more than a few miles within the floating city will undoubtedly be limited by cost considerations. The sole possibility for successful operation would be a "stringing out" of the modules with each one having its own highway link to the mainland. But there are many problems involved in tying into mainland highway facilities. (In most situations, the capacity of a highway network is determined by the intersections of the network links. It is at these points of conflict that there must be a sharing of the available space; the volumes of vehicles which can pass through these points are significantly lower than those volumes which can freely flow between intersections.)

Further, no circulation between villages or between villages and the town and city centers could be economically provided, and this lack would weaken the entire fabric of the city. For these reasons, a transportation system relying principally on the automobile is not recommended.

The second alternative is elimination of the automobile from Triton City and a reliance upon some form of public transit for circulation both within the floating community and to the mainland. Such a scheme has several drawbacks. First, it is undesirable to separate residents from their automobiles. A good example of the inconvenience is the vacation-bound resident who would have to transport all of his luggage and hardware via a transit system to his automobile on the mainland. Indeed, some of the most successful residential developments in concentrated urban areas have avoided this spatial separation through the use of on-site garaging.

The second, and perhaps more important, objection to banning motor vehicle traffic is the requirement for freight service to the floating city. At this time, there appears to be no feasible alternative to truck delivery of freight. The volume of freight to the floating city will be substantial, and any scheme which might involve a warehouse and transfer point on the mainland would be financially prohibitive.

The third system, a combination of auto and public transit, is then the most practical concept for meeting circulation requirements. The peak hour commuter movement (18,000 persons) justifies and, in fact, demands a sophisticated rapid transit system. Functionally it must connect the floating town and city centers with the existing city core, and should ideally have a loop type configuration to minimize conflicts and eliminate switching problems. This would reduce the number of stations and overall system length; at the same time it would be able to serve the movements which begin and end internally.

Freight must be moved onto the city, so it will be necessary to provide some limited highway network. To take advantage of the excess highway capacity, it is recommended that a limited amount of on-site parking be available at each village module. (The precise number of spaces is a subject for more detailed investigation.) Parking facilities will be of two types: some will be reserved for short-term parking demands such as package loading/unloading and visitor parking, and some will be provided for long-term (overnight and longer) parking of residents' vehicles.

Population densities of Triton City will be high, and the system requirements are restricted to moving people to the regional core a short distance away. The magnitude of the peak hour commuter movement, coupled with the minimum transit system length and station requirements, will create a favorable economic situation. The system can be highly automated, and the limited scope of the system itself will simplify this process The rubber tired transit expressway system as demonstrated in Pittsburgh by the Allegheny Port Authority presents one distinct possibility.

40. References:

a Wilbur Smith & Associates, Transportation and Parking for Tomorrow's Cities, pp. 286-9.

50. TECHNICAL CONSIDERATIONS

51. MARINE

Note: A general discussion of marine considerations is given in paragraphs 13. and 14., discussions of site environments.

51.1 Stability

The attitude of a floating object is determined by the interaction of the forces of weight and buoyancy. If no other forces are acting, it will settle until the force of buoyancy equals the weight, and will rotate until the centers of buoyancy and gravity are in the same vertical line. The body is said to be in stable equilibrium if, when displaced slightly in any way, it returns of itself to its original position. If, on the other hand, it moves farther from its original position when displaced slightly, it is unstable.

The ability of a floating body to remain stable under the action of overturning forces may be evaluated through the concept of the metacenter. The metacenter, with respect to any axis of rotation, is the point of intersection of a vertical line through the center of buoyancy when the body is inclined, with the original vertical through the center of buoyancy when the floating body was in level position. For any particular cross section normal to the axis of rotation, the metacenter will remain in approximately the same location for small angles of rotation, up to about seven degrees. The distance along the original vertical through the center of gravity from the center of gravity to the metacenter is called the metacentric height for that axis of rotation. The metacentric height is an index of the stability of a floating body, at least up to small angles of heel about a given axis of rotation

The metacentric height may be determined by first calculating the distance from the center of buoyancy to the metacenter, $\overline{BM}$:

$$\overline{BM_1} = \frac{I_1}{\nabla}$$

where I_1 is the moment of inertia of the water plane area about axis 1, ∇ is the volume of displacement of the floating body, and $\overline{BM_1}$ is the distance from center of buoyancy to the metacenter for axis 1.

The distance from center of buoyancy to the center of gravity, $\overline{KG}$, can be determined from the geometry and weight of a particular floating body. The metacentric height, $\overline{GM}$, can then be determined from $\overline{KG}$ and $\overline{BM_1}$ for axis 1. Whenever the metacentric height is positive, the floating object is stable against overturning when subject to small angles of heel.

The metacentric height can also be used to calculate the amount of heel caused by applied moments about axis 1 for small angles of heel.

$$\text{Moment to heel one degree} = W \cdot \overline{GM} \sin 1^{\circ}$$

where W is the weight of a floating body.

Consider Module "A" (triangular platform) as a fully floating structure when in its final location under full load. For an individual triangular unit, the metacentric height and moment to heel 1° are as follows (see Fig. 1 for location of axes):

$$\overline{BM_1} = \overline{BM_2} = 680 \text{ ft.}$$

$$\overline{GM_1} = \overline{GM_2} \approx 600 \text{ ft.}$$

Moment to heel one degree ≈ 4,300,000 ft.-kips (about axis 1 or 2)

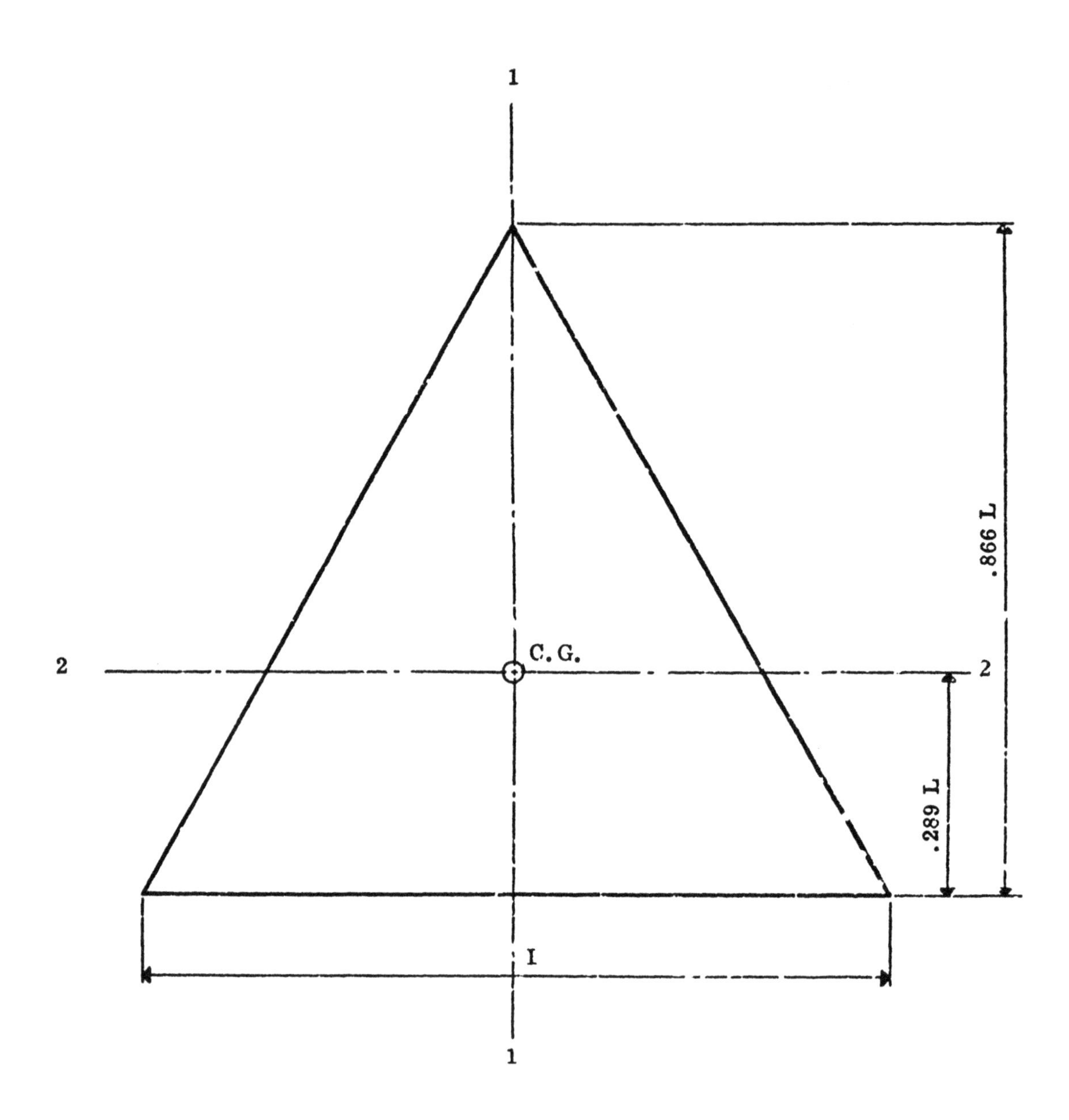

FIG. 1 - REFERENCE AXES FOR STABILITY OF TRIANGULAR PRISM

These values indicate that the triangular unit module "A", having such a large metacentric height, could not overturn. Furthermore, a wind load of 30 psf on one face of the structure causes a moment about axis 1 of approximately 400,000 ft.-kips. This moment results in an estimated heel angle of 0.1 degree or a one foot differential free-board across the structure. If all the live load which is considered to be "movable" (350 psf for the total superstructure and below decks structure) is placed on one side of the structure the expected angle of heel is about 0.86 degree or about a 10 foot differential free-board across the structure.

It seems evident from the above that the structure is stable, with a very high factor of safety for any distribution of live load, but considerable heel could develop if the maximum movable live load all occurs on one side. This theoretical possibility could not actually occur in normal functional circumstances, and any tendency to discomforting heel can be counteracted by a system of ballast (such as pumped water) to compensate automatically for moving live load and thereby to maintain the deck in a relatively level position.

A semi-floating design would not have to deal with stability and heel under movable live load, wind waves, etc. except temporarily, when the structure is under tow and in a light condition.

51.2 Response to Wave Action

Waves cause rocking and oscillation of a floating structure, as well as bending and shear effects in the hull. The height of waves and severity of wave action are functions of wind velocity and the configuration and location of the adjacent body of water. Much information is available in the technical literature on coastal engineering for estimating characteristics of waves that may be expected in various locations.[a,b,c,d] Since, in general, Triton City will be located in shallow water and in relatively protected locations, wave effects will be much less severe than in the open sea.

The significant maximum height of waves for any given site will be one of the criteria used to determine minimum free-board. The degree of tilt of the raft due to wind forces, waves, and shifts in live load will also affect the free-board requirements for a fully floating structure.

For a semi-floating structure which is supported in a fixed vertical and horizontal position, free-board will be influenced by wave heights and by the maximum tidal range. Tilt could not occur in this type of structure.

For a fully floating design, the period of roll of the raft has an important effect on comfort levels that may be anticipated in the floating structure. The period of roll about an axis 1 of a floating body subject to wave action may be estimated with the aid of the metacentric height for this axis, $\overline{GM_1}$, as follows:

$$\text{Period}_1 = \frac{1.108\, k_1}{\overline{GM_1}}$$

where k_1 is the radius of gyration of the mass of the body about axis 1 through the center of gravity of the body, with both k and $\overline{GM}$ in feet and period in seconds.

The period of roll of the Module "A" structure about either axis 1 or axis 2 is estimated to be in the range of 6 to 10 seconds. This coincides approximately with the significant period of waves expected in a large body of water of 35 foot depth during hurricane winds.

The estimated period of roll for Module "A" is rather short, compared with many ships, due to the module's large metacentric height. This means somewhat higher than normal accelerations due to rolling, and thus, lower comfort levels. Furthermore, coincidence of the wind/wave periods may increase the amplitude of roll, with further reduction of comfort levels in the structure during severe wind storms. Preliminary calculations indicate that roll may cause discomfort in a floating structure of the type under consideration here; further studies are necessary to determine if this is so, and if needed, to develop the many counter-strategies available.

51.3 Anchoring Requirements

The floating city must be anchored against lateral forces due to wind, current flow, and waves. However, the anchoring system must not prevent the substantial vertical movement due to tides and waves; restraint of vertical movement is not consistent with the philosophy of full support by flotation. Furthermore, the wide range of expected vertical movements could not be restrained without tremendous anchoring structures.

Preliminary calculations indicate the order of magnitude of lateral anchoring requirements as follows:

Wind - 30 psf	4,000,000 lbs
Current and waves	1,000,000 lbs
Total lateral anchorage required	5,000,000 lbs

The best method for lateral anchoring of the fully floating structure will depend to a large extent on the actual conditions associated with a particular site and orientation of floating units relative to land.

One of the chief advantages of the semi-floating scheme would be the virtual elimination of a need for permanent lateral anchorage. Lateral resistance would be provided by contact of the raft with the bottom, either via direct bearing on the bottom or via pile supports extended below the raft. Furthermore, no provision for free vertical movement would be required. Depending on the degree of buoyancy used for support of vertical dead lead in this scheme, it might be necessary to provide some uplift anchorage for safety against the effects of insufficient downward load to hold the structure on its bottom support

52. STRUCTURAL

52.1 General

The Triton City concept requires new applications of structural systems to provide efficient and economical structure. Several aspects of the proposed architectural solution have important structural implications:

(1) Over water location with floating or semi-floating foundation.

(2) Megastructure supporting and enclosing a number of individual standardized dwelling units within each bay of the main structure.

(3) Construction of major portion of structure in an off-site fabrication and assembly location (such as a shipyard) and water transport to the site by tow.

The above considerations describe a structure which is both a permanent, stationary building and a floating vessel like the hull of a large cargo ship. Design criteria differ from the usual criteria for buildings and for ships, acquiring some of the attributes of each. A floating structure derives its entire support from hydrostatic buoyancy, so that distribution of weight in the structure as well as wave effects and dynamic characteristics of wind storms take on far more significance than they would for an over-water building. Of course, because of the protected location of the floating city, sea and wave effects are considerably less severe than for normal ocean-going ships. Also, building requirements such as type of occupancy, fire safety and maintenance considerations will differ markedly from typical ship construction practice. Thus, the Triton City structure has its own unique design criteria, and these will not be completely representative of either a building or a ship. However, these design criteria and technical requirements for the floating city do not extend beyond the present state-of-the-art in the fields of building construction and ship construction.

An evaluation of the feasibility of floating foundations is important in order to compare the cost of the floating system with conventional foundation construction at the same over-water site. Conventional systems include the following types:

(1) Bulkhead and land fill with the structure supported on piles and reinforced concrete pile caps.

(2) Piles standing free in water with reinforced concrete pile caps.

(3) Deep poured-in-place basement and mat foundation on bottom soil of suitable bearing capacity.

The optimum conventional system would vary with site conditions, but system (2) above would be a contender at most sites. This system is used herein for comparison with several floating or semi-floating systems.

The following general criteria provide a basis for development of structural systems for the module superstructures:

(1) Typical building code requirements will govern the construction and no substantial variances from loading, structural or fireproofing requirements will be necessary.

(2) The foundation must be a floating or semi-floating design. Therefore, the weight of the structural frame and fireproofing are important characteristics of the system, and must be minimized to the greatest extent consistent with the best economics of the combined superstructure and base structure

(3) The bay size of the apartment and core area must be large enough to permit considerable flexibility of space without disturbing the basic megastructure framing.

(4) The framing system must be capable of being prefabricated using current shipyard technology.

(5) Modifications of the basic frame or local strengthening should be possible.

52.2 Code Requirements

Most of the provisions of local building codes will apply to Triton City. While specific provisions vary somewhat among cities in different areas of the country, the design requirements listed are typical of some of the widely recognized codes.[e,f] In the final design, the criteria of appropriate building codes would normally be used.

The "Rules and Regulations of the U.S. Coast Guard for Passenger Vessels, CG256,"[g] contain specific instructions on fire protection, stability, subdivision into watertight compartments and fire zones. Informal discussions with local Coast Guard officers indicate that these regulations are probably not jurisdictionally applicable to Triton City as long as it is permanently anchored to the shore.

The following live load provisions are typical of most building code requirements in all parts of the country:

Apartment floors	40 lbs/sq.ft.
School classrooms	50 lbs/sq.ft.
Walkways and public spaces	100 lbs/sq.ft.
Commercial space (supermarket)	125 lbs/sq.ft.
Access roads (truck loading)	250 lbs/sq.ft.
Parking (automobiles only)	60 lbs/sq.ft.
Parking (automobiles and trucks)	250 lbs/sq.ft.

The above loads are maximum intensities which may occur over relatively small areas. For structural members and elements which support relatively large areas (such as lower story columns and foundations), the above loads may be greatly reduced. For a structure the size of module "A", most codes permit a reduction of 50% to 60%[h] to be applied to apartment and school room design live loads for lower columns and foundations. Design live loads for public and commercial spaces are not reduced.

In addition to live loads which result from occupancy of the building, the following typical ranges of environmental loads must be considered:

Snow load on roofs:	20-40*lbs/sq.ft.
Wind on vertical profile:	20-35 lbs/sq.ft.
Earthquake:	.05 to .15 times building dead load

* May be higher for unusual locations in extreme northern climates.

Both structural steel and reinforced concrete construction will be used in various parts of the project. Most building codes now recognize the following national specifications which have been developed for structural steel and reinforced concrete:

(1) Structural Steel--American Institute of Steel Construction, "Specifications for the Design, Fabrication and Erection of Structural Steel for Buildings"

(2) Reinforced Concrete--American Concrete Institute, "Building Code Requirements for Reinforced Concrete" (ACI 318-63)

In addition to the above standards, the specifications of the American Bureau of Ships will be used for materials and welding practices relative to the floating base structures.[i]

52.3 Protection: Fire and Flooding

Typical building codes limit the floor area of incombustible framing between firewalls to 5,000 to 7,500 sq.ft., and the number of stories of incombustible framing between firewall separations to three. Firewall enclosures must have a rating of 2-3 hours; columns and girders: 4 hours; and floor systems: 3 hours

Some of the bulkheading and fire protection techniques and criteria developed for shipboard use which meet CG256 might be adaptable to Triton City after sufficient testing.

Passenger ships are required to be subdivided into watertight compartments and provided with a double bottom to prevent a disaster in the event of puncture of the hull by collision or running aground. While the problem of Triton City running aground could be solved by adequate anchorage and preparation of the harbor bottom under and near the structure, the possibility of collision could exist in some locations. Therefore, measures to positively prevent structural damage from a collision or to limit the amount of flooding in the event of a break in the hull will have to be provided if Triton City derives its entire support from hydrostatic buoyancy,

52.4 Life of Structure, Maintenance, and Deterioration

The design life of the structure should be a minimum of 50 years. Expense for maintenance of the base structure is undersirable since conventional foundation construction generally does not require maintenance. If maintenance costs are anticipated, their capitalized value must be included in the economic evaluation of the proposed base structure.

Protection against corrosion will be needed for a steel base structure. Polluted sea water is one of the most corrosive environments for steel. Unlike ocean-going passenger ships, which are drydocked each year to permit inspection of the hull and replacement of fouled or corroded hull plates, this structure must be inspected in place and protected from corrosion indefinitely. Because of the complex nature of the corrosion problem, occurring with or without the presence of oxygen; the variety and ever changing nature of the pollutants in a typical salt water harbor; and the special problem near the water line, several types of corrosion protective systems will have to be employed for a steel hull. The primary protection would be provided by a cathodic system used in conjunction with organic coatings, or concrete or gunite mortar. Furthermore, periodic inspection of both inside and outside of the hull will be required. Corrosion of steel at the water line and in the splash zone above the water line will not be inhibited by cathodic protection.[j] Here, a concrete or gunite encasement would be used to protect the steel.

There is considerable experimental work and field experience to demonstrate that air-entrained, dense concrete made with selected aggregates and cement will perform satisfactorily in corrosive sea water conditions for a number of years without supplementary protection systems.[k to q] Galvanized reinforcing steel may be used for increased protection against corrosion of reinforcing. There is evidence that concrete made from lightweight aggregate may also give satisfactory performance in sea water. A concrete ship, built in 1919 using lightweight aggregate and later partially sunk in a harbor, was examined in 1953; the concrete and reinforcing steel were reported to be in good condition.[r]

52.5 Superstructure

Design Loads: Design live loads are governed by local building codes. Typical values are given in paragraph 52.2. Design dead loads are as follows:

Apartment Area:	Weight of prefabricated apartment units	4.0 lbs/cu.ft
Core Area:	Allowance for partitions	20 lbs/sq.ft
All Areas:	Structural steel and fireproofing weight	as computed

Architectural and functional considerations determine many of the basic dimensions of the structural system. The main constraints are as follows:

Apartment Areas: Vertical supports are to be spaced at 33.3 ft. on center along each side of the main triangle. No longitudinal cross bracing is permitted in the space between vertical supports. Horizontal supports in the megastructure are to be spaced at 27 ft. to permit three tiers of prefabricated units within each unit of megastructure. Each set of three prefabricated units is to be supported on the floor structure and vertical frames of the megastructure. Prefabricated units may be omitted in various locations. The transverse dimension of the typical apartment unit is 50 feet.

Core Area: Functional requirements are compatible with a pattern of columns at the vertices of 50 ft. equilateral triangles. Height between levels is 27 feet. For certain occupancies, a mezzanine floor is added between the basic levels of the megastructure. Steel floor beams arranged in a cartesian system are supported on steel trusses located along the edges of 50 ft. equilateral triangles. Columns are at the vertices of the triangular arrangement. A 5 inch thick structural lightweight concrete floor slab is used in this area.

Fire Protection: The necessary fire protection of steel beams, columns, and girders is a lightweight, hard-setting spray-on coating which is recognized and accepted by the Underwriters Laboratory.[s,t]

Horizontal fire barriers in the apartment areas are provided at each primary floor (27 ft. o.c.) by 2.5 in. of lightweight concrete cast on metal decking, which has its underside protected with the spray-on fireproofing mentioned above. In the core area, the 5 in. thick structural, lightweight concrete slabs supported on fireproofed steel give the necessary fire protection.

Vertical fire separation in the apartment areas is provided by a system of insulated panels.

52.6 Base Structure

The base structure is, in effect, the foundation of the building above. The estimated loads which are imposed on the base from the superstructure (order-of-magnitude estimated from first preliminary design stage) are given in paragraphs 52.2 and 52.5. Also in this section are additional loads imposed from the main floor and three lower floors of the base structure which may be used for parking, mechanical equipment, or storage.

For a fully floating structure, all loads are supported by buoyancy of the base structure or raft. The supporting reaction is uniformly distributed when the raft is of uniform depth and floating in a level position in still water. To avoid major bending and shear effects in the raft, applied loads must also be distributed over the base structure in exactly the same manner as the reactions developed by buoyancy. Since, in general, this is not possible in any practical structure which supports movable live load, the base structure will require considerable bending and shear strength.

Furthermore, waves cause a nonuniform distribution of hydrostatic buoyancy. The distribution of support reaction varies as the wave crest moves across or along the floating structure. As a result of these wave effects, additional bending and shear strength is required in the raft for support of applied load with any type of distribution. Foundation support by true flotation is best suited to structures in still water locations with their principal loads distributed in close accord with the distribution of the buoyancy.

A "semi-floating" concept may be used to reduce these overall bending and shear requirements and to gain other advantages without the loss of most of the functional benefits of a floating structure. In this approach, the base structure is designed as a floating raft for support of its own weight and that part of the superstructure which is assembled prior to its movement to the site. The site is prepared in advance to take a portion of the dead load and all of the live load which the structure will eventually support. After partial fabrication in an off-site plant location, the structure is towed to the site. Once in proper position, it is either lowered to uniform bearing on a prepared bottom bearing layer, or piles are extended downward for support on, or below, the bottom.

Applied Loads: The total weight estimate for the superstructure above the top deck of the base raft is 97,000 tons. This is broken down as follows:

	Load (Short Tons)	
Apartments- dead load	51,000	
Interior structure - dead load	8,000	
Stair and elevator towers - dead load	3,000	
Earth on top deck of raft	4,000	
Total superstructure dead load		66,000 tons
Apartments - live load*	17,000	
Interior structure - live load	3,000	
Stair and elevator towers - live load	1,000	
Top deck of raft - live load	10,000	
Total superstructure live load		31,000 tons
Total superstructure - dead and live load above deck		97,000 tons

Additional loads for the structure below deck are the anticipated two parking decks, a mechanical equipment area within the hull, and a storage deck. The total weight estimate for below deck loads is 36,000 tons. This is broken down as follows:

2 Parking decks - dead load	10,000	
Mechanical equipment	3,000	
Total dead load - below deck floors		13,000 tons

* Live loads are "reduced live load" see paragraph 52.2.

Brought forward - total dead load - below deck floors		13,000 tons
2 Parking decks - live load	7,000	
Storage deck - live load	16,000	
Total live load - below deck floors		23,000 tons
Total below deck floors - dead and live loads		36,000 tons

The estimated total superimposed load which the raft must support is 133,000 tons (short tons). Of this total weight, dead load is 79,000 tons and live load is 54,000 tons. The average unit loading is:

Dead load	830 psf	
Live load	560 psf	
Total load		1,390 psf

The weight of the raft itself must be added to the above loads to obtain the total displacement of the structure.

52.7 Existing Technology: Major Flotation Structures

Few permanent buildings are supported by flotation or by partial flotation. Undoubtedly, there have been a number of good reasons for this: cost of raft compared to conventional foundation construction, dynamic effects of wind and sea, corrosion of metal and concrete in sea water or fresh water exposures, and anchoring difficulties. Nevertheless, some significant permanent structures are supported by full or partial floating; and of course, large ships, some over a thousand feet long, are actually floating structures. The Lake Washington floating bridge is a major permanent structure supported by flotation. Pier 57 in New York City is a semi-floating structure, partially supported by buoyancy. Sections of the structure were fabricated some 38 miles up the Hudson River and towed to their final location next to Manhattan Island. The BART Subway tunnel tubes for the San Francisco Bay crossing were fabricated at an assembly facility some distance from the site and towed to their final location where they were sunk in a prepared trench on the bottom.

Pier 57 consists of three large, reinforced concrete boxes. The buoyancy of the boxes supports a large portion of the dead weight of the pier. The remainder of the dead load and all the live load is supported by bearing on the bottom and by a small number of composite steel and concrete piles which also act as dowels to anchor the substructure. The two larger boxes are 150 ft. wide by 360 ft. long by 32 ft.-6 in. deep and weigh about 27,000 tons each. Unit weight is approximately 46 lbs per cubic foot.[u, s]

The BART tube sections are steel shells approximately 47 ft. wide by 366 ft. maximum length by 24 ft. depth. The tube structure consists of a 3/8 in. thick exterior steel shell which is exposed to severely corrosive soils and derives its resistance to corrosion from cathodic protection.

Typically, large ships are of welded steel construction, with dimensions ranging up to 1,000 ft. long by 150 ft. wide by 50 ft. draft. Unit weight of hull structure is in the range of 6 to 10 pounds per cubic foot of hull volume.[x] Highly automated construction methods, which utilize a high degree of prefabrication and sophisticated equipment for handling and assembly, have been developed, particularly in Japan and Europe, for construction of certain common types of tankers and cargo vessels.

Large ships for ocean and coastal service were constructed of reinforced concrete in World Wars I and II. The remains of a reinforced concrete (lightweight aggregate) tanker which went into service in 1919 and which was sunk a few years later at the entrance to Galveston Harbor, off the Texas coast, were examined in 1953 during an investigation sponsored by the Expanded Shale Institute. This ship was a 75,000 ton vessel with dimensions: 434 ft. long by 54 ft. wide by 26 ft. draft, with 4 inch thick sides and a 5 inch bottom. Samples of concrete taken out in 1953 and tested for compressive strength indicated strengths in excess of the original design strength. Both concrete and reinforcing steel were in good condition after some 34 years of continual exposure to sea water. The approximate unit weight of the hull was 31 lbs per cubic foot, of which about 7 lbs per cubic foot was reinforcing steel.[r] The ratio of the weight of the vessel to the weight of cargo was about 0.58, compared to a typical value of about 0.20 for a steel tanker.

52.8 Alternative Base Structures

Several alternatives have been considered for the raft structure. The three concepts which appear to have sufficient merit for further study are:

Scheme 1 - Separate steel hull units under apartment structures with independent raft for support of central core

The principal base structure is made up of three steel hulls, one along each side of the main triangle under the apartment areas. Each hull has dimensions of approximately 116 ft. x 667 ft. x 30 ft. maximum draft. These hulls are connected at the corners to form a rigid equilateral triangle with an open central core.

The central core area is supported on an independent floating raft of triangular shape, 264 ft. on a side. This core may float at a different level than the edge hulls under certain conditions of live load or wave action. Links between the two sections must offer adequate flexibility to compensate for differentials of several feet between adjacent deck levels of hull and raft.

The type of construction proposed is similar to modern tanker construction. The weight of steel required is determined by the local pressure, transverse span and longitudinal bending moment. Local pressure and transverse spans in this scheme are about the same as for a modern super tanker of some 40,000 ton cargo displacement. Maximum longitudinal bending moment is probably of the order of one third to one half of the hull girder bending moment used for a similar size ocean-going tanker (order of magnitude 8,000 to 12,000 ft-kip per ft. of width for Triton raft). On this basis, the estimated weight of steel plates and siffeners in the hull and main deck is about 6 lbs per cubic foot of base volume or about 210 lbs per sq.ft. for a 35 foot deep base.

The steel would be protected by pneumatically applied cement mortar, epoxy–coal tar coatings, cathodic protection, or a combination of several methods. Pneumatic mortar (or concrete on the main deck) might add some 80 psf to the above hull weight.

With this scheme, the estimated total weight of the raft structure is about 27,000 tons, giving a total weight of fully loaded raft and superstructure of 160,000 tons.

Scheme 2 - Single triangular steel raft

This approach utilizes a single triangular raft, 667 feet on a side. The raft is framed in steel with a series of internal trusses connecting a steel upper deck to the steel bottom. The main floor and the bottom act as flanges, and the trusses act as a two-way web system to make the overall raft behave as a rigid plate structure capable of carrying the required shear and bending effects which are the results of movable live load and waves. The raft unit bending moment is estimated to be of the same order of magnitude as for the hull structures of Scheme 1 Hence, the estimated unit weight and requirements for corrosion protection are about the same as in Scheme 1.

The internal trusses needed for adequate structural rigidity might impose undesirable restrictions on the use of the interior space for parking. Ballast is required in the central portion of the raft for a better match of dead load and buoyancy distribution. Since the central portion is an integral part of the rigid raft, there are no differences in level of flotation between central and edge regions as may occur in the first approach.

Scheme 3 - Concrete edge hull and independent raft for core support

A third approach is to construct the base as a reinforced concrete structure in a manner similar to Scheme 1. The unit weight of this system can vary between about 30 lbs. per cu.ft. for a design with high strength, lightweight concrete, and 45 lbs per cu.ft. for a design with stone concrete. Thus, for a 35 foot total depth of the base, the estimated weight range for the base is 1,000 psf to 1,600 psf, giving a total weight of base structure of 96,000 tons to 154,000 tons. This includes the weight of the main floor, the two intermediate parking floors, the intermediate storage floor, the bottom, the exterior sides and all interior bulkheads. The weight of fully loaded raft and superstructure in this scheme will range from 216,000 tons to 274,000 tons.

The base structure, with either lightweight concrete or stone concrete, is much heavier than a steel base. This is a serious disadvantage if it is desired to support the entire live and dead load by flotation. On the other hand, the additional weight of a concrete base, compared to steel, might not be a disadvantage for a structure which must only float while being towed from the fabricating yard to the site. In this case, the base will support by full flotation only a portion of the superstructure dead load plus the base dead load. Additional superstructure dead load and all live load can be supported by the earth below the structure when it is in its final location. A concrete base can provide the necessary buoyancy with a draft of less than 30 feet and with considerably less bending strength requirements than for the fully floating system.

Lower maintenance and much better corrosion resistance are decided advantages of the concrete system. Proper selection of cement, aggregates, and mix design can further enhance the suitability of concrete for this application.

53. MECHANICAL

53.1 Heating, Ventilating & Air Conditioning

The compact size of Triton City modules means that all piping and ductwork for distribution systems can be held to minimum lengths, thus reducing costs and space required for installation. In addition, the spatial organization of each module, i.e., residential areas grouped around the outside of the module and common usage spaces clustered in the interior, accommodates the most efficient distribution systems for all mechanical services. Due to the high population densities, there are fewer external surfaces than in conventional developments, and operating loads and costs for heating, cooling and ventilating are greatly decreased. Since the loads are reduced, the required equipment is also smaller and takes less space and less capital outlay than it would in a more dispersed development.

The immediately available supply of ocean or fresh water to be used in condensing equipment for the mechanical services obviates the need for cooling towers and will further cut cost and space required, and eliminate a major eyesore in traditional construction. The large surrounding body of water also offers a means of eliminating toxic fumes and undesirable odors from the ventilating and other exhaust systems, so that there is no danger of air pollution. Thus, the central open space in the village module can be a large outdoor or enclosed volume containing completely fresh and unpolluted air.

Distribution systems for these services will be centralized so that an economical source of clean, tempered and treated air can be provided, particularly in an urban area of high air pollution, for it is cheaper to treat and clean a large volume of fresh air through central equipment than it would be if there were individual systems for each of the various functions within the neighborhood modules. Residential air handling equipment will be installed at main distribution branches with individual temperature and humidity controls in each apartment. Air handling equipment for the common usage spaces, e.g. schools, stores and offices, will be located within each specific area. There are two advantages to this: first, it gives each function individual control over the climatic conditions within its space and, second, it keeps the distribution system to minimal size.

Approximately 23,000 sq.ft. will be required in each module of 5,000 persons for central plant mechanical equipment. This equipment will all be housed in part of one of the subsurface levels of the megastructure. Risers for all mechanical distribution--including plumbing and electrical--can be accommodated in approximately 2,250 sq.ft. (or about 750 sq.ft. in each of the three vertical towers containing elevators and located at the corners of the village module "A"). Horizontal or "branch" distribution for all systems can be run along the underside of the public walkways and will require about two feet of depth, except in areas of concentrated usage where three feet may be required.

53.2 Plumbing

Depending upon the site of the floating city and the capacity of the existing facilities within the metropolitan region, Triton City may use the metropolitan water and sewage systems or may eventually develop its own. It is probable that any initial community would connect to the facilities of the core city. This connection will require flexible junctions to allow for movement of the floating platforms, but otherwise does not present any technical problems. With the water sources that are available to Triton City, it is entirely possible that a floating city of 100,000 persons would have its own water and sewage disposal system. In this manner, it could aid in the general drive toward depollution of water resources and could, by providing an additional source of potable water, help to alleviate metropolitan problems of water shortage during dry seasons.

As in the case of other mechanical equipment, plumbing for the Triton City modules gains from close, highly organized disposition of functions. Again this means that plumbing distribution will be very efficient, and both central equipment--water heaters, chillers and pumps--and individual piping runs can be of a smaller size and a lower cost than in usual developments. Space required for central plumbing equipment and distribution systems has been accounted for in paragraph 53.1., above. The only unique feature of the distribution system, distinguishing it from plumbing in ordinary buildings, is the fact that the pitch of horizontal piping runs will have to be 1/4 inch per foot instead of the usual 1/8 inch per foot. This is necessary so that, when the platform is tilted slightly under wave action, the piping will still retain a downward slope.

53.3 Electrical

The compact Triton City design permits economical distribution systems for all electrical services, particularly the primary electrical system. Feeder conduits for power wiring, lighting, communications, and other low-voltage control wiring can easily be installed and serviced in the vertical riser shafts already mentioned. Connections to apartment and commercial areas can be made conveniently from these main distribution points. The concentrated load resulting from dense settlement makes possible a dual-primary electrical system. This means that there are two separate power supplies to the floating modules, so that if one should fail, the other will still be operational.

Geographic location of the city will determine the best primary and/or secondary energy sources. Generally, a total energy plant would increase capital cost for the city; however, operational savings might compensate for this, depending on local fuel and energy costs. Total electrical energy requirements for a village of 5,000 population are:

18,000 KVA Electric A.C.

10,000 KVA Gas A.C.

(Schematic arrangement for a possible total energy plant for Triton City is shown in paragraph 53.5.)

53.4 Equipment Weight (for 5,000 population)

Central plant equipment, including main piping and water at the equipment:	930,000 lbs
Auxiliary equipment, piping, conduit and ductwork @ 450 lbs./person	2,250,000
Approximate weight of water in main distribution systems	1,600,000
Total weight:	4,780,000 lbs

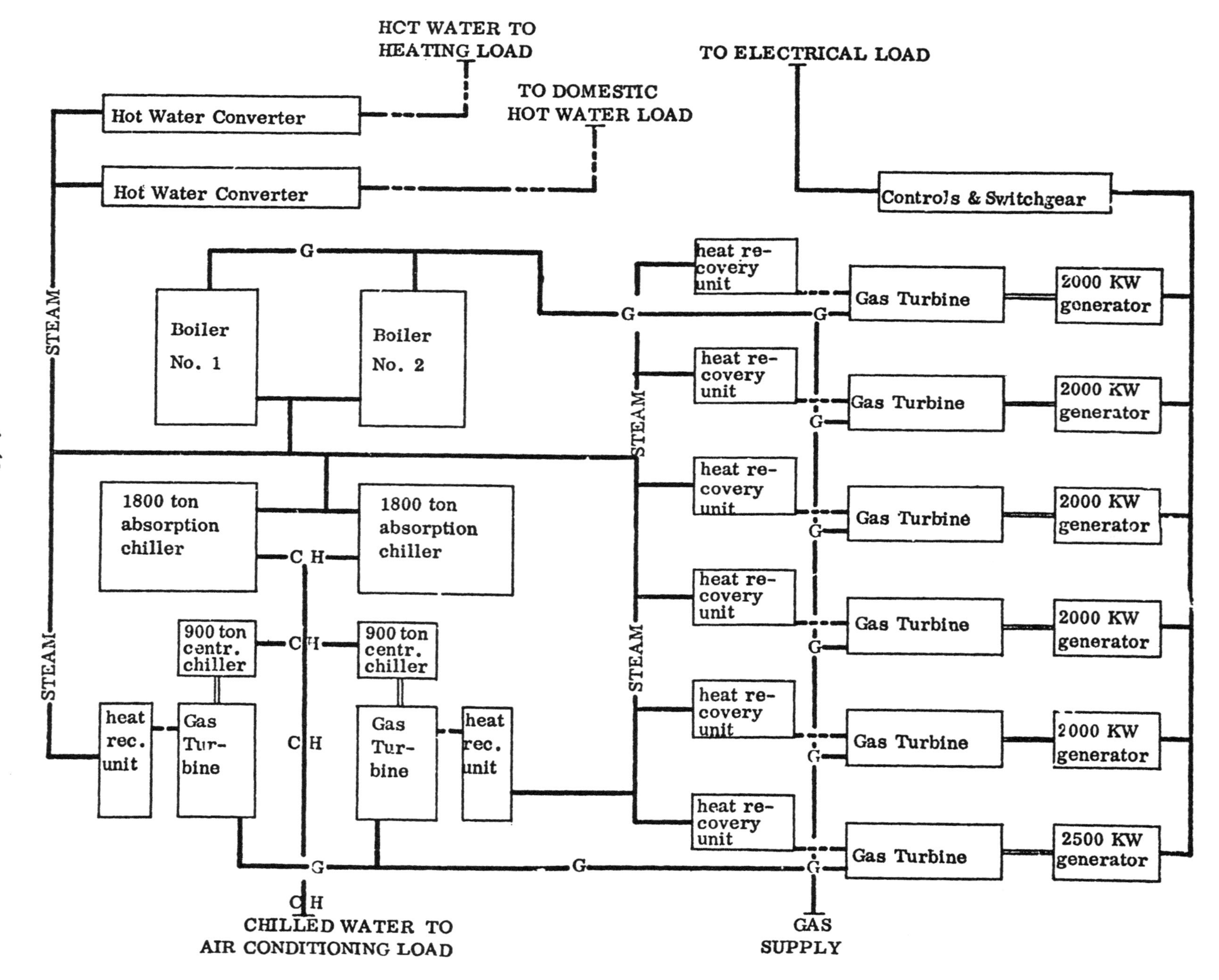
HCT WATER TO HEATING LOAD
TO DOMESTIC HOT WATER LOAD
TO ELECTRICAL LOAD
Hot Water Converter
Hot Water Converter
Controls & Switchgear
Boiler No. 1
Boiler No. 2
heat recovery unit
Gas Turbine
2000 KW generator
heat recovery unit
Gas Turbine
2000 KW generator
heat recovery unit
Gas Turbine
2000 KW generator
heat recovery unit
Gas Turbine
2000 KW generator
heat recovery unit
Gas Turbine
2000 KW generator
heat recovery unit
Gas Turbine
2500 KW generator
STEAM
STEAM
STEAM
STEAM
1800 ton absorption chiller
1800 ton absorption chiller
900 ton centr. chiller
900 ton centr. chiller
heat rec. unit
Gas Turbine
Gas Turbine
heat rec. unit
C H
G
CHILLED WATER TO AIR CONDITIONING LOAD
GAS SUPPLY

54. CONSTRUCTION

54.1 Construction Industry

The United States has not yet developed a housing industry comparable in scale and efficiency to advanced segments of the economy such as metals, automobile fabrication or petrochemicals. The industry remains a high cost, custom operation largely in the hands of small entrepreneurs, hampered by obsolete local building codes and neolithic unions. By the latter part of the 1960-1970 decade a substantial number of experimental projects aimed at converting home building into a modern, mass production industry are in the planning stage and a few projects have reached execution. As compared to the immediate postwar decade (1945-1955), relatively few of these projects involve construction of detached single family units. Following the ill-fated postwar (1945-1955) attempts to produce low cost housing in wartime factories, the market was gradually reclaimed by the traditional builders. There were, however, three exceptions to this trend. The first was the success of a handful of large-scale, new town developers like the Levitt Brothers. The second was the growth of the shell home industry. And the third was the great expansion of mobile home production; by the mid-1960's, one out of every seven new homes built in the nation was a mobile home[aa], produced by techniques more closely resembling manufacture of automobiles than the piecework, house building industry.

The focus in the late 1960's seems to have shifted, or perhaps more accurately broadened, to include high-rise structures. Experiments in the renovation of older apartment buildings have been attempted, and some appear to have borne out Vernon's gloomy prophecy[bb] that full-blown modernization of old high-rise units is as expensive as building new homes in the suburbs.

The emphasis in rebuilding cities with the help of lower cost housing is increasingly turning toward new high-rise structures. Experiments with precast concrete, with steel constructions and with other materials and techniques to arrive at lower cost efficiencies through various combinations of component fabrication and on-site assemblage are underway in various parts of the nation. Proposals calling for enlargements of the mall principle to encompass entire new towns have also been advanced, and the notion of "cities under glass" no longer meets with instant ridicule.[cc]

54.2. Off Site Plant Fabrication

Triton City represents a blending of two of these themes; the proposed megastructure calls for a unique, integral combination of new towns and new high-rise buildings. Moreover, in its own way, the proposal offers partial solutions to unanswered urban questions by developing a new composite of proven techniques outside of the constraints of the existing system of building codes, financing, construction unions, and a craft-scale of operations which has fettered the construction industry. Just as the great growth in mobile home production was attained in part by by passing the barriers which retard advances in traditional home building, the floating city can serve as a useful break with an obsolescent past and present.

Among the potential beneficiaries of this method of unified new town construction may be some unexpected entries. For example, the world's shipbuilding industry has completed super tankers of substantially more tonnage than the proposed floating villages. Over and above the possibility of utilizing some of the latent potential of the shipyards for production of large, prefabricated units, there would be major participation for traditional construction-oriented industries, notably steel and steel fabrication.

The city would also have the advantage of assembling sufficient volume in a single systems project to attract first-rate contractual and management capability and to employ efficient technology on a large scale.

The potential for introduction of new technology, greater automation and a more integrated approach to the entire building system appears best if a major portion of the building is fabricated and assembled in a plant. An over-water site in an urban location offers the possible opportunity for water transport of the entire preassembled structure or substantial sections of it.

54.3 Materials and Construction

Materials and construction for the major framework are discussed in paragraph 52.5, Superstructure. This "megastructure" of fireproofed steel is the main framework for the entire structure, within which components will be placed. Because of the quality control inherent in the shop fabrication techniques used, dimensional stability can be held to a much closer tolerance than is possible for traditional building construction. This means that sub-units of construction which are also shop fabricated can be simply installed within the main frame at minimum expense. [It is this economic approach that will allow for custom modifications at very little additional cost. These are available for our automobiles and should be for our homes.]

Materials for the units will be mostly lightweight metal alloys which are easy to work, dimensionally stable and durable, yielding long building life. Then, by economizing upon the basic construction, amenities such as adequate acoustical treatment, custom finished interiors, and special features can be incorporated into each individual unit. Similarly, the centralized provision of services and utilities allows both for additional space within each unit and for economies in equipment, since the individual residences need be fitted only with plugs, receptacles and outlets for the services instead of the actual equipment.

This type of construction has proven itself in every area where it has been used. It is cheaper initially (lower capital cost) and requires less maintenance (lower operating cost) than the traditional building construction by which we are presently limited. When these advantages are combined with the unique opportunities for individual selection of materials, room layouts and space utilization and treatments, it becomes evident just how many of the luxuries currently missing in everyday living can be realized in the floating city scheme.

50. REFERENCES

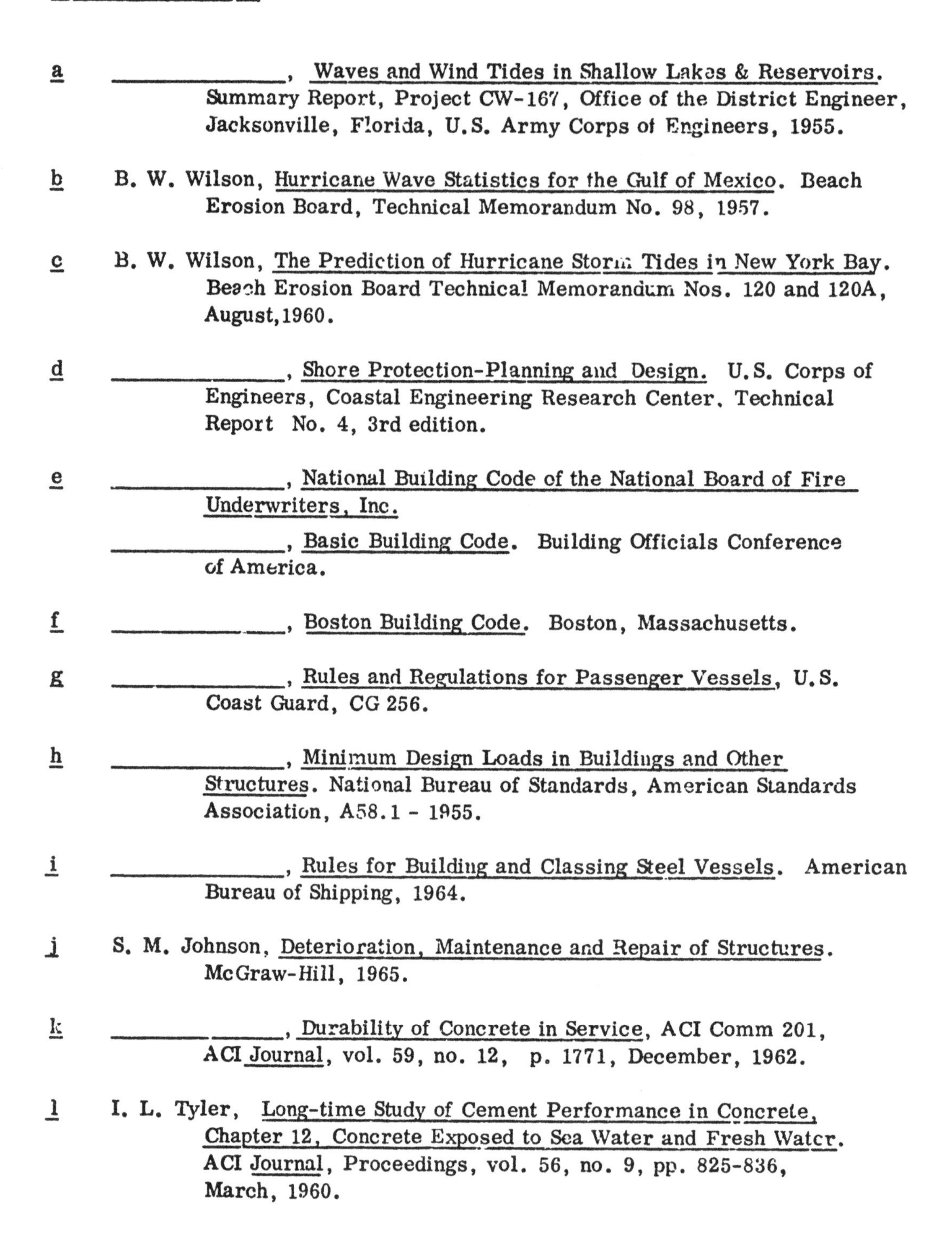

a ______________, Waves and Wind Tides in Shallow Lakes & Reservoirs. Summary Report, Project CW-167, Office of the District Engineer, Jacksonville, Florida, U.S. Army Corps of Engineers, 1955.

b B. W. Wilson, Hurricane Wave Statistics for the Gulf of Mexico. Beach Erosion Board, Technical Memorandum No. 98, 1957.

c B. W. Wilson, The Prediction of Hurricane Storm Tides in New York Bay. Beach Erosion Board Technical Memorandum Nos. 120 and 120A, August, 1960.

d ______________, Shore Protection-Planning and Design. U.S. Corps of Engineers, Coastal Engineering Research Center, Technical Report No. 4, 3rd edition.

e ______________, National Building Code of the National Board of Fire Underwriters, Inc.

______________, Basic Building Code. Building Officials Conference of America.

f ______________, Boston Building Code. Boston, Massachusetts.

g ______________, Rules and Regulations for Passenger Vessels, U.S. Coast Guard, CG 256.

h ______________, Minimum Design Loads in Buildings and Other Structures. National Bureau of Standards, American Standards Association, A58.1 - 1955.

i ______________, Rules for Building and Classing Steel Vessels. American Bureau of Shipping, 1964.

j S. M. Johnson, Deterioration, Maintenance and Repair of Structures. McGraw-Hill, 1965.

k ______________, Durability of Concrete in Service, ACI Comm 201, ACI Journal, vol. 59, no. 12, p. 1771, December, 1962.

l I. L. Tyler, Long-time Study of Cement Performance in Concrete, Chapter 12, Concrete Exposed to Sea Water and Fresh Water. ACI Journal, Proceedings, vol. 56, no. 9, pp. 825-836, March, 1960.

m B. Tremper, J. L. Beaton, and R. F. Stratfull, Corrosion of Reinforcing Steel and Repair of Concrete in a Marine Environment, Part II. Bulletin 182, Highway Research Board, 1958, pp. 18-41.

n Bryant Mather, Factors Affecting Durability of Concrete in Coastal Structures. Technical Memorandum no. 96, Beach Erosion Board, June, 1957.

o S. Halstead and L. A. Woodworth, The Deterioration of Reinforced Concrete Structures under Coastal Conditions. Transactions, South African Institution of Civil Engineers, vol. 5, no. 4, pp. 115-134, April, 1955.

p F. M. Lea, and N. Davey, "The Deterioration of Concrete in Structures", Journal of the Institution of Civil Engineers, vol. 32, p. 248 et seq., 1949.

q K. A. Adamchik, "Causes and Prevention of Deterioration of Concrete in Marine Structures in the Zone of Fluctuating Water Level." Corrosion of Concrete and its Prevention, Academy of Sciences of the USSR, published by the Israel Program for Scientific Translations.

r ______________, The Report of an Investigation on the Condition and Physical Properties of Expanded Shale Reinforced Concrete After 34 Years Exposure to Sea Water. Technical Problems Committee, Expanded Shale Institute, Washington, D.C., 1953.

s ______________, Building Materials List published by Underwriters Laboratories, Inc.

t ______________, Published data of the Zonolite Division of W. R. Grace & Company.

u ______________, "New York's Pier 57 Founded on Two 27,000 Ton Reinforced Concrete Boxes." Civil Engineering, March, 1952.

v ______________, "Pier 57 Foundation Box Towed to Final Site," Engineering News Record, 7 August 1952.

w G. J. Murphy, and D. N. Tanner, "The BART Trans-Bay Tube " Civil Engineering, December, 1966.

x J. P. Comstock (editor), Principles of Naval Architecture, Society of Naval Architects and Marine Engineers, New York, 1967.

y F. N. Spellar, Corrosion: Causes and Prevention. Mc-Graw Hill, 1951.

z L. M. Applegate, Cathodic Protection. Mc-Graw Hill, 1960.

aa _______________, Statistical Abstract of the United States, 1967. Tables 1081 and 1088, pp. 721 and 724. Data are for 1966. 217,000 mobile homes were produced in 1966.

bb Raymond Vernon, Metropolis 1985. Harvard University Press, Cambridge Massachusetts, 1960.

cc _______________, "People Under Glass". Wall Street Journal, November 27, 1967, p. 20.

The gingerly favorable editorial refers to the HUD-financed Minnesota study of a $6 billion glass-domed city of 250,000 persons.

60. COSTS AND COST COMPARISONS

Initial cost estimates suggest that the expense of building floating communities compares favorably with that of conventional construction on land. Both current trends and predictions indicate that the entire fabric can be provided for about $8,000 per person. This includes housing, schools and other community facilities, all services, roads and utilities (including air conditioning), as shown below.

61. Triton City Preliminary Cost Estimate

Neighborhood Unit Cost

Raft (includes storage parking & mech. space)	$ 7,000,000
Megastructure frame & floors	3,000,000
Vertical circulation	1,750,000
Mechanical services (HVAC, plumbing, fire protection, water sewage)	12,600,000
Apartment units	8,000,000
School 72,000 sq.ft. @ $15/sq.ft.	1,000,000
Architectural[1]: stair towers, side rails, landscape, glazing, etc.	3,000,000
Total cost of neighborhood	$ 35,850,000

Cost per inhabitant of neighborhood $7,200.

---

[1] Excluded are architectural costs of commercial and other revenue producing spaces.

Extra infrastructure (located in town and city center) per neighborhood

Highways allow 300 feet per unit @ $600 per lin.ft.		$ 180,000
High school 185,000 sq.ft. @ $25 per sq.ft. = $4,620,000 ÷ 4		1,250,000
Government 15,000 sq.ft. @ $25 per sq.ft.		350,000
Per person cost $400	Total:	$ 1,780,000
allow 100% contingency	Total: $800 per person	

Approximate total cost per person is:

Neighborhood cost	$ 7,200
Other	800
Total Cost	$ 8,000

62. Preliminary Cost Breakdowns:

Raft

193,000 sq.ft. @ 230 lbs./sq.ft. (36' depth @ 6.3 lbs/cu.ft.)[1] 22,000 tons @ $210	$ 4,620,000
Protection 250,000 sq.ft. @ $1.60	1,000,000
Parking floors 260,000 sq.ft. @ $4.00	400,000
Anchorage allowance	1,000,000
	$ 7,020,000

[1] Information from Dravo, Inc.

Megastructure frame

Steel: apartment + corridors
1,400,000 sq. ft. @ 8.5 psf = 12,000,000#
115,000 sq. ft. @ 20 psf = 2,300,000#
14,300,000# @ $12.50/# = $ 1,800,000

2 1/2" conc. over 250,000 sq.ft. @ $1.20/sq.ft.	300,000
5 1/2" conc. over 115,000 sq.ft. @ $1.25/sq.ft.	150,000
Spray on fireproof 2,750,000 sq.ft. @ $.15/sq.ft.	425,000
	2,675,000
Contingency	325,000
	$ 3,000,000

Vertical circulation[1]

24 passenger elevators @ 60,000 each	$ 1,450,000
6 freight elevators @ 50,000 each	300,000
	$ 1,750,000

Mechanical

Sewers	410,000		
Water	205,000		
Electric power	(625,000)[2]		
Telephone	(250,000)[2]	(1,490,000)	$ 615,000[2]
HVAC	6,800,000		
Plumbing	1,850,000		
Fire	590,000		
Electric & Communications distribution	2,900,000	(12,140,000)	$ 11,485,000[3]

[1] Information from Otis Elevator Co.

[2] Costs will be borne by franchised and public utilities, cf. Columbia, Md Leisure Life, N.J. and Reston, Va.

Information from American Telephone & Telegraph Co.

[3] Figures given include commercial revenue producing areas; therefore, $655,000 has been deducted from totals to exclude costs not included in other comparative schemes.

Total mechanical $ 12,100,000

Apartment units

1,000,000 sq.ft. @ $8.00 $ 8,000,000

This $8,000 per person can be measured against estimates of $7,500 per person for suburban developments (without many of the urban institutions, availability of existing public transit, or air conditioning) and over $11,000 per person for renewed urban areas (still without air conditioning and other amenities) (See paragraph 63.).

63. Costs per Person: Conventional Construction

It is extremely difficult to assemble a comprehensive set of cost figures for the entire fabric of a community or city. In the case of existing cities, growth and development have been incremental; few if any records were kept of original costs, and even where these were recorded, it is very hard to assess contemporary valuation. This adds to the problems of calculating renewal expenses. How much of the existing city must be included in the cost for renewal of any small area within that city? Moreover, many of the "hidden costs", e.g. relocation, are almost impossible to determine with any accuracy; and there is no consistent method for evaluating costs since various portions of a single renewal program are handled by different agencies, each of which has its own evaluation techniques

In the case of new towns, some of the cost problems of cities are eliminated, but there still remains the difficulty of ascertaining total cost from many contributing agencies and individuals. Many of the expenses in such new developments are never compiled or made public.

It is within this context that the following cost comparisons have been derived, and both the accuracy and the comprehensiveness of many of the figures can be legitimately questioned. However, it may be stated that refinements in these estimates would _raise_ rather than lower the costs.

63. Cost Comparisons: Conventional Construction

FACILITY	EXISTING CITIES	NEW TOWNS - U. S. A.		NEW TOWNS - FOREIGN			CITY EXPANSION		URBAN RENEWAL	
	Cambridge, Mass.[a]	H. H. F. A. Proposal [b]	Annadale-Hugeunot, N. Y. [c]	Tama, Japan Proposal [d]	Basildon, England[e]	Bracknell, England [f]	Camden, N. J.[g]	Isard Costs [h]	St. Louis, Missouri[i]	Isaacs Costs [j]
RESIDENTIAL	4000.00 (est.)	4533.-	4500.- (est.)					?	5720.-	1740.-
INSTITUTIONAL	304.44	423.-	1171.-							
Schools	211.00	223.-	724.-					430.0		142.-
Other	93.44	200.-	447.-					?		75.30
GOVERNMENT	81.04	?	286.-					?		37.50
COMMERCIAL	?	563.-	?					?		917.-
INDUSTRIAL	?	133.-	?					?		1375.-
RECREATIONAL & OPEN SPACE	11.22	incl. in inst.	incl. in gov't.					?		82.-
SERVICES & UTILITIES	266.00 + water	200.-	2428.-					75.-		74.20
STREETS & ROADS	152.53	50.-	684.-					145.-		75.-
LAND	1500.00 (est.)	160.-	1100.- (est.)					?		included
Total: (Date)	$6,624.- (1955)	$6,485. (1965)	$11,240.- (1966)	$3,366.- (1965)	$2,330.- (1965)	$2,200.- (1965)	$10,000.- (1965)	$650.- (1956)	$5,720.- (1967)	$4,523.- (1961)
Adjusted to Equivalent American Costs [k]	—	—	—	x2.5	x3.0	x3.0				
Adjusted to 1967 Costs[l]	x 1.67	x 1.15	x 1.08	x 1.15	x 1.15	x 1.15	x 1.15	x 1.54	--	x 1.25 x2[6]
TOTALS:	$11,080.[1]	$7,450.[1]	$12,150.[1]	$9,680.-[2]	$8,040.-[3]	$7,590.-[3]	$11,150.[3]	--[4]	$5,720[5]	$11,300.-[4]

1 Incomplete total.

2 Does not include air conditioning, kitchen equipment, vertical transportation, etc.

3 Does not include cost of land.

4 Insufficient data.

5 Residential costs only.

6 Actual renewal was for 1/2 of population: adjusted total reflects this. New and rehabilitated housing was for less than 1/4 of metropolitan area population so cost/person could be considered as over $24,000.

64. Cost Comparisons: Cost/Person vs. Population Density

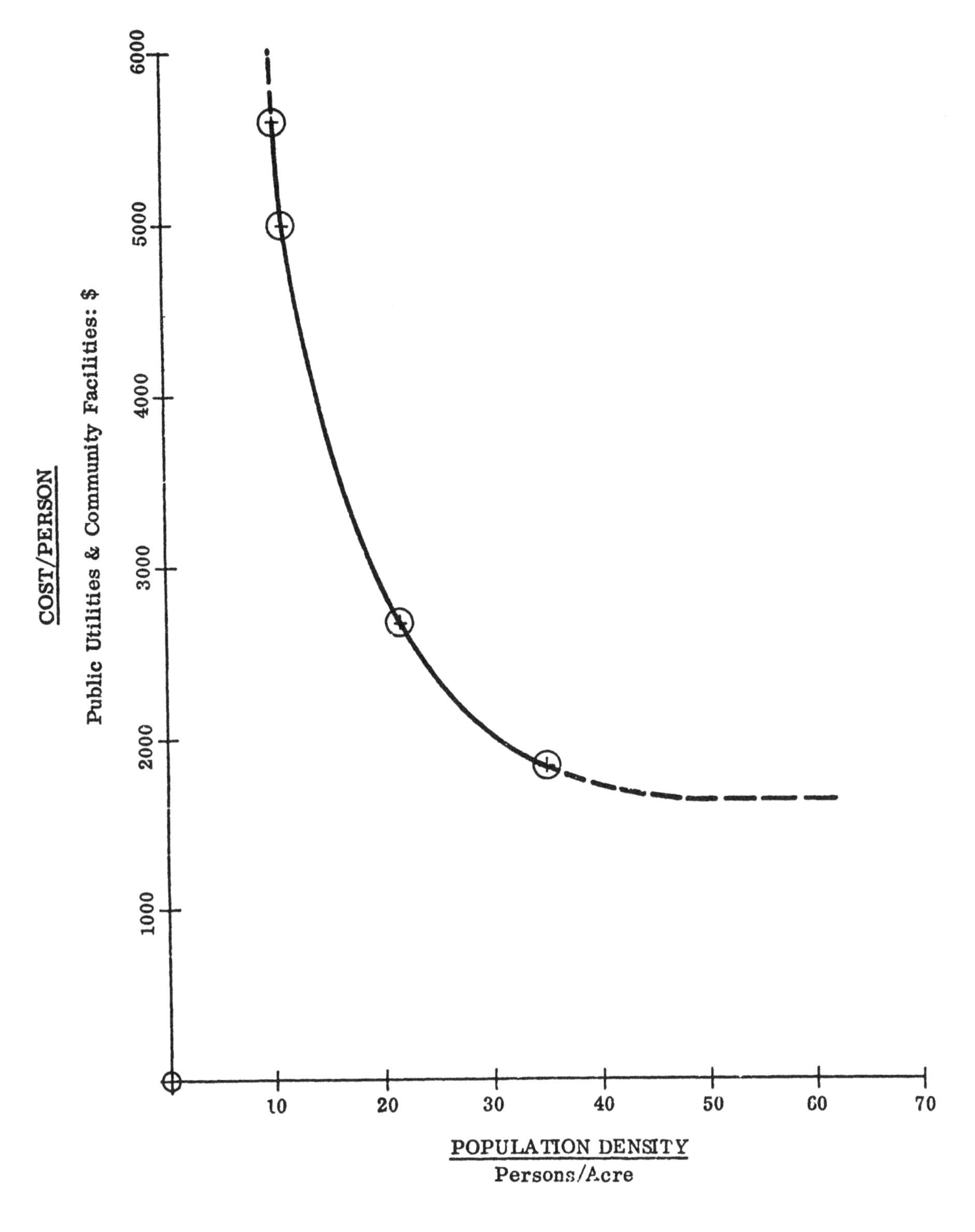

SOURCE: Ref.[c]

65. Land and Renewal Costs

Cost figures for renewal depend not only upon section of the country, density of the population, and location within the city, but also upon the nature and quality of buildings in the area or neighborhood under consideration and the type of work to be done. The table below lists some typical ranges:

ESTIMATED COSTS:[m]	% OF TOTAL COST			
	Survey, Plan Administration	Property Acquisition	Demolition & Clearance	Site Im-provement
Clearance Areas:	4.34	78.6	9.15	7.86
	1.65	77.9	12.7	7.79
	4.07	70.3	20.2	7.03
Residential Conservation Areas:	16.7	38.6	10.1	3.84
	13.8	70.6	8.6	7.1
	22.3	63.7	7.4	6.3
	100.	--	--	--
	13.4	48.2	33.6	4.8
	15.5	65.8	12.1	6.5
Arrested Development Areas:	19.2	37.7	3.4	39.9
	19.4	55.6	3.2	21.7
	7.5	61.3	11.3	19.7
	20.8	28.4	2.1	50.5
	8.7	24.6	5.1	25.9
	17.7	22.6	4.6	20.8
	29.3	23.2	6.3	41.-

The cost of land acquisition for urban renewal can be approximated at between 40 to 60 % of the total project cost.[n] Moreover, cleared site costs of $15 to $100 per square foot (higher for prime sites) are no longer unusual in large cities. The cost of land and associated buildings in portions of New York City is as high as $500 per square foot.[c] At this rate, the floating city modules would be extremely inexpensive when compared to costs in the adjacent urban core. But, in fairness, they probably should not be compared with the $500 to $5,000 per acre costs of the properties assembled for new towns in suburban and exurban areas.[b]

60. References:

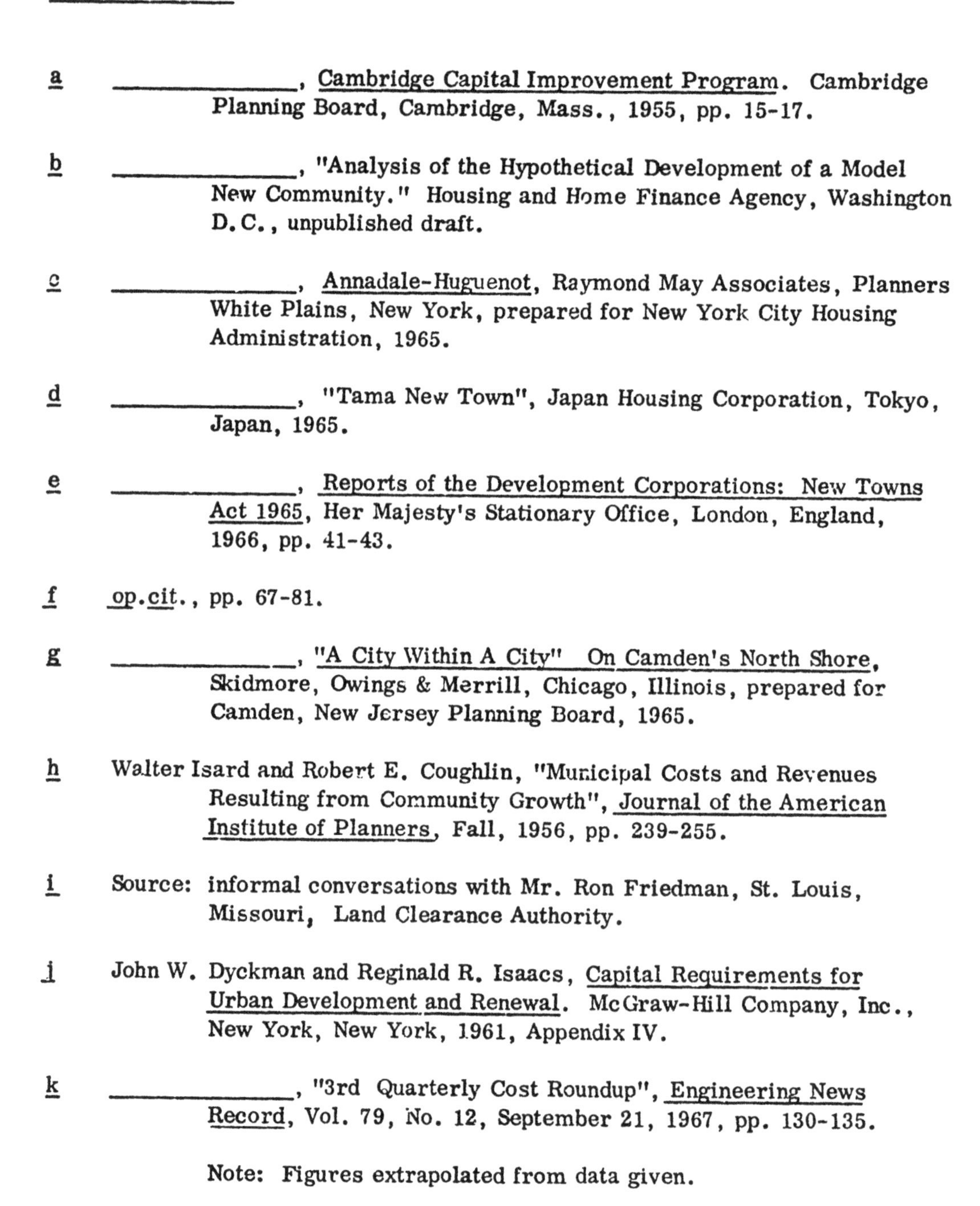

a ______________, Cambridge Capital Improvement Program. Cambridge Planning Board, Cambridge, Mass., 1955, pp. 15-17.

b ______________, "Analysis of the Hypothetical Development of a Model New Community." Housing and Home Finance Agency, Washington D.C., unpublished draft.

c ______________, Annadale-Huguenot, Raymond May Associates, Planners White Plains, New York, prepared for New York City Housing Administration, 1965.

d ______________, "Tama New Town", Japan Housing Corporation, Tokyo, Japan, 1965.

e ______________, Reports of the Development Corporations: New Towns Act 1965, Her Majesty's Stationary Office, London, England, 1966, pp. 41-43.

f op.cit., pp. 67-81.

g ______________, "A City Within A City" On Camden's North Shore, Skidmore, Owings & Merrill, Chicago, Illinois, prepared for Camden, New Jersey Planning Board, 1965.

h Walter Isard and Robert E. Coughlin, "Municipal Costs and Revenues Resulting from Community Growth", Journal of the American Institute of Planners, Fall, 1956, pp. 239-255.

i Source: informal conversations with Mr. Ron Friedman, St. Louis, Missouri, Land Clearance Authority.

j John W. Dyckman and Reginald R. Isaacs, Capital Requirements for Urban Development and Renewal. McGraw-Hill Company, Inc., New York, New York, 1961, Appendix IV.

k ______________, "3rd Quarterly Cost Roundup", Engineering News Record, Vol. 79, No. 12, September 21, 1967, pp. 130-135.

Note: Figures extrapolated from data given.

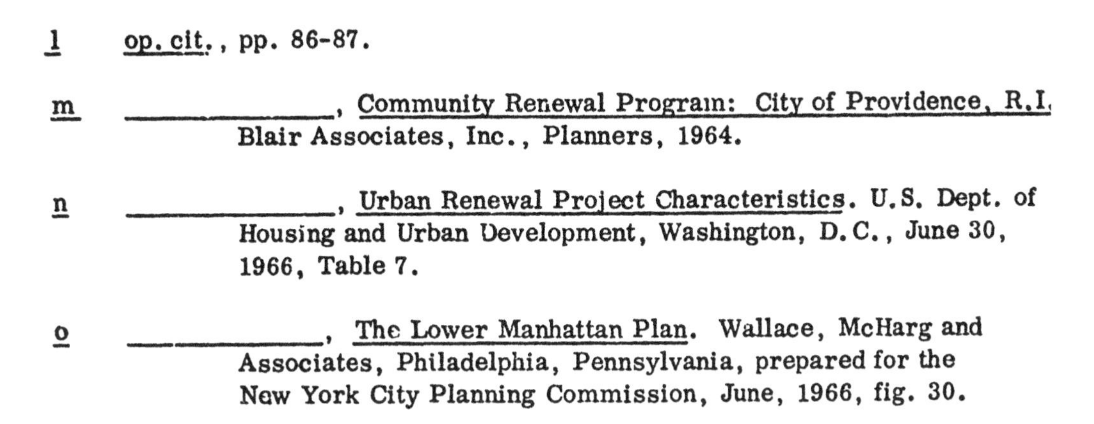

l op. cit., pp. 86-87.

m ______________, Community Renewal Program: City of Providence, R.I. Blair Associates, Inc., Planners, 1964.

n ______________, Urban Renewal Project Characteristics. U.S. Dept. of Housing and Urban Development, Washington, D.C., June 30, 1966, Table 7.

o ______________, The Lower Manhattan Plan. Wallace, McHarg and Associates, Philadelphia, Pennsylvania, prepared for the New York City Planning Commission, June, 1966, fig. 30.

70. CITY PROBLEMS AND TRENDS

71. City Problems

The downtown areas of our cities are suffering from the problems created by a combination of pressures: at the same time that metropolitan population is multiplying, the buildings, transportation systems and all other services--in short, the whole fabric--of the declining central cities are in need of major rehabilitation and upgrading. Many of the structures, residential, commercial and institutional, are old, dilapidated, and do not function efficiently to meet contemporary needs. Land is scarce, expensive, and so densely used that there is, quite literally, no room to move.

To further complicate the problem, population in general, and metropolitan area population in particular, has increased at a very high rate and will continue to do so. Expansion, therefore, is mandatory, for these rising numbers of people must be accommodated.

Generally, there are two kinds of urban expansion: horizontal and vertical. In the 1940's and 50's, most expansion was horizontal, i.e., city dwellers moved out into the suburbs, producing what is now known as "urban sprawl." This exodus from the core area into outlying communities which are primarily residential has had certain substantial disadvantages. Many people who live in the suburbs work in the city and must get in and out of town, thereby taxing anachronistic transit systems or creating monumental traffic jams in the downtown area, or both. Additionally, the suburbs of large cities are a drain on municipal facilities, most of which are probably outmoded to begin with. Also, and very importantly, suburban communities have drawn a large number of industries and persons in the middle and top income levels, thereby depleting the city's economic base and leaving it peopled with low income families, many of whom require intensive public services.

Vertical expansion has taken place, to a certain extent, in densely settled cities through the construction of multi-story commercial, institutional and apartment buildings. Sometimes sizable building sites are "discovered" by reconverting military land to civilian use, by reuse of rail yards, by filling shore areas, or more frequently, by assembling parcels devoted to "lower uses" to some higher use. The last technique, often used in connection with urban renewal, has been violently attacked as a form of "Negro removal" or an assault on the inadequate supply of low income housing units available to the poor. Expanding in this direction almost always means using land which is already committed and involves all the economic and political problems of acquisition and demolition of existing structures and relocation of the people in them.

Traditional horizontal expansion has only served to increase service demands on the city center while depriving it of much of its tax revenue. Such vertical expansion as has been accomplished has been so tremendously expensive, due to the high price of urban land, relocation costs, and the general expense of traditional construction, as to be prohibitive as a complete solution to the problem.

It is possible, however, to expand both vertically and horizontally and, in fact, to build new high density communities immediately adjoining the urban core without consuming land and without disturbing people or neighborhoods. Many large cities are located on bodies of water (river, lake, bay or harbor) and, in most cases, the water is the only uncommitted open space near the central area. The technological capability now exists to construct whole new villages, towns and cities and float them off-shore of urban centers.

72. Tradition of Land Fill

We tend to assume that all sections of our large cities have always been there; but this is not the case, for as they grew and became crowded, expansion was often accomplished by means of land fill into the water, thus creating additional land close to the city core. Boston, Massachusetts, where much of the current downtown area and large portions of the in-town residential neighborhoods are built on "filled land", is well known for this.[a] Other major cities have equally striking examples of filling in harbors or swampy areas or extending shorelines out into rivers or lakes. For example, the new Candlestick Point in San Francisco is filled land, and many of the famous Florida Keys are completely man-made land. The recently completed study for Lower Manhattan Island proposes pushing the shoreline of the lower island out to the ends of the existing docks. When this occurs, this portion of the island will have doubled its original size.[b]

Extension into surrounding water area may be accomplished by filling in the water, as most of our cities have done, or by floating rafts, boats or other platforms on the water surface, as has been the tradition in the more crowded coastal cities in Asia, most notably Hong Kong.

73. Water Recreation Statistics

Our society is experiencing a newly burgeoning interest in the water surrounding or adjoining the land masses which we have traditionally thought of as our natural domicile. This enthusiasm manifests itself as a result of increased affluence and leisure time and shows itself in the growing numbers of people who crowd to the beaches, lakes and rivers for vacations and relaxation. Boating, fishing, swimming and other water sports have seen a tremendous growth, to the point that boating, once a pasttime only for the well-to-do, is now so common that many "average" families have their own boats. Similarly, the number of summer cabins, cottages and campsites at water recreation sites has expanded very rapidly in the past few years.

Visits to National Forests[c]	1950	1965	% Increase
Swimming (1000)	902	3,749	415%
Fishing (1000)	4,885	19,358	396%
Camping (1000)	1,534	10,420	680%
No. of Fishing Licenses Sold[c] (1000)	15,338	24,472	160%
Money Spent on Sports Equipment,[c] Boats, Etc. ($ million)	869	2,996	345%
No. of Outboard Motors in Use[d] (1000)	2,811	6,784	242%
No. of Boats Sold[d] (1000)	131	298	228%

74. Waterfront Living

Part of the heightened interest in water is a new concern with urban waterfront areas. A number of major cities have begun renewal campaigns for their hitherto badly neglected waterfront properties. Most notable for this planning are the eastern seaboard cities of Baltimore, Boston, New York, and Philadelphia. Old wharfs and piers are being reclaimed for use as commercial and recreational sites. Warehouses are being converted to luxury apartments. In Boston, several of the old wharfs which have already been rehabilitated are so successful that there is a two year waiting list for vacancies.[e] The extent of this movement is such that in many large cities the most promising area is just that waterfront which a few short years ago had been one of the worst problems.

Rehabilitation of existing structures is just one of the possibilities open to cities. In Camden, New Jersey, directly across the Delaware River from Philadelphia, a new residential development, directly on the waterfront, is planned for 15,000 people. The land for this development is largely old city-owned dump and derelict property.[f] The Lower Manhattan Plan, recently completed for New York City, proposes the addition of 40,000 new dwelling units for that area, with almost three-quarters being waterfront residences--accommodations for nearly 100,000 persons.[b] Other major cities have similar proposals and plans in various stages of progress, but just as the number of existing waterfront structures is limited, so is the available shoreline property.

75. Advantages of Floating Cities

Triton City can offer significant built-in advantages of water and open space amenity and can function either as an aid to urban renewal processes, as a means of helping to accommodate expanding population, or as an independent satellite town, adjacent to but distinct from the city center.

One of the great advantages of the floating city is that it entails no disruption of the existing fabric of the city or loss of dwellings. The new town would add to rather than diminish the supply of housing, augment the number of viable neighborhoods, and provide additional facilities to the urban core. It is also conceivable that the movement of city residents into the new community would free for renewal the land on which they previously lived.

The technique is similar to land fill in that it creates new, inexpensive land near the city core, but floating facilities have the added benefit of flexibility since they can be moved to new locations should conditions dictate.

1. It permits metropolitan expansion as close as possible to the city core, thus minimizing wasteful and costly travel time and traffic congestion.
2. Temporary or permanent relocation for city dwellers can be effected so that subsequent renewal of the core can be undertaken.

3. By eliminating the high cost of land acquisition from the total expense, more amenities for low cost housing and urban construction can be economically provided.

4. These amenities themselves can offer many social benefits, e.g., air-conditioned houses for summer comfort will reduce "street loitering"; adequate lighting of public areas can help to depress the crime rate; proximity of water recreation and other open space will allow an outlet for the expenditure of other frustrated energies.

5. By employing mass production techniques for the construction efficiencies for conversion of use and maintenance, as well as cost, can be realized.

6. Since the units will float, they can be moved to comply with future demands and changes in city requirements.

7. Capitalizing upon this technique could have unique advantages resulting from new and different concepts of ownership, city management and resident responsibility.

70. References:

a ______________, 1965/1975 General Plan for the City of Boston and the Regional Core. Boston Redevelopment Authority, Boston Massachusetts, 1965, Chapter II.

b Wallace, McKay & Associates, et alia, The Lower Manhattan Plan, prepared for The New York City Planning Commission, New York, June, 1966.

c ______________, Pocket Databook, 1966. U.S. Bureau of Census Washington, D.C., 1967.

d ______________, Statistical Abstract of the United States, 1965. U. S. Bureau of Census, Washington, D.C., 86th ed., 1965.

e Source: Boston Redevelopment Authority, Boston, Massachusetts, various publications.

f Skidmore, Owings, & Merrill, "A City Within A City" On Camden's North Shore, prepared for Camden Planning Board, Camden, New Jersey, October, 1965.

80. SOCIAL AND ECONOMIC CONSIDERATIONS

81. Preliminary Considerations

The proposed construction of 5,000 population floating villages adjacent to cities appears to offer a number of meaningful advantages, including partial solutions to some widely identified urban problems. It is important, however, to recognize at the outset that the new towns, which may comprise only a fraction of one percent of metropolitan area population, cannot in themselves be expected to solve all of the accumulated social ills of our urban areas. The floating facilities can make a useful, limited contribution to alleviation of critical problems if they are not allowed to fall victim to the wracking internal conflicts which beset most larger cities.

While Triton City is applicable to many "unique" requirements of location, industry, or neighborhood groupings, this study is developed in the context of a satellite community with a balanced population and a fairly high proportion, of residential use. In addition, some initial screening of population, commerce and other activity may be desirable in order to provide a control group which can furnish data for planning subsequent developments. For example, will the inclusion of a very good school in the neighborhood module have significant effects on population characteristics and modes of behavior?

82. Human Scale

Criticisms of the overwhelming size, the impersonality, and the personal alienation characteristic of life in large urban aggregations have been growing in frequency in recent years. Partial solutions to this problem include "new towns in-town" as recommended by Perloff;[a] satellite cities on the British model; or simply conservation and fostering of neighborhood communality either through spontaneous regeneration as suggested by Jane Jacobs[b] or through Alinsky-type community protest organization.[c]

It may be noted that the Perloff proposal calls for a transformation of urban gray areas into viable communities by energizing the total environment for population sectors of 50,000 to 100,000. The new towns would include a "Lighted Center" offering lively night time activity; perhaps a sunken plaza; good quality public schools; and a focus on providing job opportunities partly through the creation of nearby industrial estates.

The floating villages offer a relatively attractive means of developing neighbor hoods and human scale in a partially controlled setting. In achieving this objective, there are obvious guideposts and warnings of past errors (ranging from the dull, soulless public housing projects to the almost equally dull suburban subdivisions) which should not be ignored.

83. Safety in the Streets

One of the principal reasons for the desertion of cities by middle class residents and for the complaints of much of the remaining population is the rising level of urban crime and violence. Triton City offers an opportunity to attack this problem in part by efforts to restrict potential sources of socially pathological behavior. It is conceivable that techniques can be devised to avoid rentals to persons with significant police records, a history of drug addiction, or severe mental instability. Violence from roving gangs can be minimized by police surveillance at the connecting causeway(s). It may well be that most of the crimes against persons are committed by relatives or close friends,[d] but statistics are often less meaningful than emotions. In many larger cities, entire areas are considered unsafe by virtue of well-publicized attacks by gangs and muggers.

As suggested by Jane Jacobs,[e] effective street lighting, the presence of human activity, and the absence of project dead spaces attractive to barbarians can undoubtedly help to curb this type of criminal activity and thereby encourage a sense of security in residents. Moreover, there is much to be said for attempting to shun the fact and appearance of beleagured citadel characteristic of many core area oases inhabited or used by middle class people. Nevertheless, investigation may disclose that the stockade approach is not entirely wtihout merit, particularly during periods when the city is undergoing a particularly painful social transition in which violence or the threat of violence is endemic.

84. A Quality School System

Under present circumstances there is no guarantee that any large urban school system can continue to offer quality education. The loss of highly motivated and well-prepared children to the suburbs and an influx of poorly motivated and poorly prepared children from disadvantaged families have placed an enormous burden on urban schools. Efforts to accelerate educational progress by busing to eliminate racial imbalance have had only minor success, and the avenues of improvement now seem to lie in other directions. These include compensatory education, including curriculum enrichment, in slum schools; smaller class sizes; community involvement; and the construction of large educational parks, the latter to offer remediation and schooling for the disadvantaged and a presumably attractive alternative to suburban schools for middle income families.

Triton City will draw families with school age children to the extent that it can provide schools which compare favorably with those in the suburbs. This probably depends in part on the degree to which it is permitted to restrict residents to the relatively sober, industrious middle class and poor. If Hylan Lewis[f] is correct in his estimate that roughly half the poor are stable and well-motivated in terms of middle class standards, then rent supplements, income supplements or other means can be implemented to evolve a community and a school system which include a broad range of income groups but are not shattered by the presence of large numbers of problem children.

Two other factors must also be considered. The first is the independence of the Triton City schools. It is possible that, through busing or other approaches the community will be drawn into the maelstrom of conflict in big city school systems. One of the prior arrangements which deserves attention in the planning and location phases is the floating city's relationship to such onshore issues as the future of neighborhood schools. The degree of autonomy to be accorded in the design and operation of the Triton City schools should be ascertained. This autonomy is important in its influence on a major factor in determining school quality: the ability of the teaching and administrative staff. A new town with a good student body located close to an urban center can expect to attract able personnel only if it is allowed some independence in school affairs. Obviously, this depends on a resolution of the sensitive question of school control.

85. Reducing the Journey to Work

Over the years, there has been some dispute between planners and planning agencies and highway engineers over the proper relationship between residences and work places. Planners often feel that a reduction in the journey to work by means of public transportation linking compact residential areas to work places is a desirable objective. In contrast, highway engineers contend that the future lies in adaptation to the popularly demanded pattern of dispersion. They suggest that the best transportation system is one which frankly recognizes dispersion and diffusion as the trends of the future and which expands the range of job choices through an effective area road system.

Whatever the relative merits of these schools of thought in meeting metropolitan transportation needs, there appears to be no doubt that a substantial number of urban residents either prefer to live in town near their place of work or are constrained to do so by reason of income. Assuming that a key objective of Triton City is to retain middle income families in the urban core, one of its chief attractions is minimization of the journey to work. It is conceivable that a suburban resident employed in the core city can spend as much as an hour to two hours in door-to-door travel. If the floating city is so located as to offer a home to work journey of perhaps half an hour to forty five minutes, it could provide a benefit to the employees of five to ten hours a week or 2,500 to 5,000 hours over a ten year span. This is surely one of the major advantages of the floating community and must clearly weigh as a significant factor in any cost-benefit calculation.

At the present time, it is known that many urban residents are willing to pay a substantial premium to live in town.[g] The market for potential middle income, in-town renters includes a number of single persons, married couples with pre-school or post-school age children, and some families with school age children who either can afford private schools or who have reason to believe that their neighborhood schools are of reasonably good quality. There are often tradeoffs between the attractions of the city, including the abbreviated journey to work, and the lure of suburban schools, greenery and public safety. A new town which offers certain suburban qualities, along with a relatively short and painless trip to work place and city entertainment, should possess holding power for middle income families.

86. Uglification and the Automobile

Ranking high on the list of most disliked features of large cities are their ugliness, noise, danger from automobiles, odors and dirtiness. There are those who charge that this adds up to a concerted effort to drive out residents who possess even a modicum of sensitivity and the income necessary to relocate

Triton City would present a viable alternative to in-town living. Usable open space; absence of automobile dangers on the streets and exhaust fouling the atmosphere; a clean, attractive environment; well-maintained buildings and neighborhoods would all be available a short distance from the work, educational and entertainment centers of the city. Abrogating automobiles to a subsurface level would in itself add immeasurably to the desirability of the community.

It must be recognized, however, that the floating city is not a fully autonomous entity. In close proximity to the city, it inevitably shares some of its dirt, visual squalor and air pollutants. Siting the community so as to minimize the deleterious aspects of this proximity is obviously an important criterion in the preliminary planning.

80: References:

a Harvey Perloff, ". . . . .", Journal of the American Institute of Planners, May, 1966.

b Jane Jacobs, "Unslumming and Slumming", The Death and Life of Great American Cities. Random House, New York, 1961, pp. 270-290

c Saul D. Alinsky, Reveille for Radicals. University of Chicago Press, Chicago, 1946.

NB: Alinsky has been active in helping to organize population groups in slum areas, including Polish-Americans, Pureto Ricans and Negroes on the basis of opposition to the "establishment," City Hall and the "interests". Alienated minorities are mobilized to demand better jobs, better treatment in housing, schools and retail stores.

d Marvin E. Wolfgang, "Urban Crime", The Metropolitan Enigma, James Q. Wilson, ed., Chamber of Commerce of the United States, Washington, D.C., 1967, pp. 253-255.

e Jane Jacobs, op.cit., pp. 40-44.

f Hylan Lewis, Poverty's Children. A Project Conducted by Cross-Tell, Washington, D.C., September, 1966.

g Charles Abrams, The City is the Frontier. Harper and Row, New York, 1965, pp. 116-123.

NB: Abrams indicates that the "blessings of good site location" have resulted in a good market for the $65-$70 per room monthly apartment rentals in parts of Manhattan, including Greenwich Village and Washington Square North.

90. RECOMMENDATIONS AND SCOPE OF FURTHER STUDY

91. Social, Economic and Educational

Some preliminary social and economic considerations associated with Triton City have already been identified. In the next study phase, it will be necessary to pursue certain key topics including the following:

(1) Relations with City Governments. Whether linked to cities as temporary experiments or permanent settlements, new floating communities raise a number of intricate legal and political issues. Tax policy, purchase of municipal services, and the degree of local autonomy over rentals, schools and police are only a few of the questions needing further study. Are the villages to be operated by independent corporations contracting for services with the municipal government? Will the management be given wide latitude in screening tenants and/or visitors? How much jurisdiction will city school systems, police departments and, indeed, trade unions of public and private employees have in the new towns?

(2) Alternative Uses and Mixes. It has been suggested that the proposed megastructure can accept a variety of uses. These can range from the balanced. primarily residential community described to a unit predominantly devoted to office buildings, medical facilities, public education, or specialized types of industry. Defining and costing the market for potential uses on various scales (e.g. one to twenty village modules) and relating the results to the needs of various communities (and community sub-areas) is a necessary study task.

(3) Social Implications and Roles. The roles of the floating villages in relieving municipal pressures and meeting municipal goals and the probable internal impact--and stress--entailed by high density socio-economic mixes, deserve serious attention. The relations among market factors; social decisions; and the physical aspects of the villages including, for example, organizational patterns on a community and sub-community level are also extremely important. Instant loyalty and morale in a new community are certainly achievable but can be gained nly by means of careful planning.

(4) Receptivity Study. There are definite dangera that any substantial new development may fall victim to clashing interest groups. Advocates for the Negroes and the poors spokesmen for craft and other unions, and for municipal agencies and agency employees, and for private real estate owners each have objectives which may diverge from the primary goal of creating a viable community. The next study phase must identify the conditions and assurances which are prerequisites for successful location.

(5) Ownership and Rental Patterns. The new communities will require detailed analysis of the nature, extent and desirability of alternative ownership and rental patterns for residential and other types of space. In part, the choice and mix of patterns may be influenced by the scale of the project and by the income levels and personal preferences of the residents. The degree to which the villages are conceived as experimental operations, with a maximum number of options to be kept open for future decision, is also a significant guiding factor.

92. Transportation

Any transportation system for Triton City should utilize the megastructure in order to reduce construction costs. This produces problems for a fixed-rail type transit system, since relative differences in elevation due to wave and tidal movements must be overcome. The magnitude of these differences need further investigation, and the feasibility of compensating for them should be more thoroughly explored.

It has already been noted that the organization of the floating modules will have great bearing on both the efficiency and the feasibility of the transportation system. A considerable amount of work will be necessary to establish the most favorable module arrangement in terms of transportation requirements.

Another consideration is the environment in which the transportation system will operate. This is particularly important in northern climates where conditions of termperature and humidity may be more severe.

93. Planning and Organization

Subsequent studies will encompass more detailed planning on all levels of the Triton City scheme--from the design of alternative apartment layouts and configurations, through specific uses of generally allocated spaces within the town and city centers, to comprehensive design of alternative arrangements for the floating city itself and its possible relationships and linkages to the existing city. Circulation and movement of both pedestrians and vehicles will be examined in depth, and the results will be one of the determinants to be incorporated in the next design phase.

As part of the further investigation of city configuration, some specific locations will be selected and recommended as choices for initial implementation. Alternative arrangements of facilities and growth patterns will be detailed in this phase of activity

Additionally, very careful attention will be given to the individual dwelling units themselves, in terms of the type, the size, and the utilization of space within the units (including materials, surface treatments, special equipment, environmental controls and living patterns) and also in terms of the possible relationships of apartments to adjacent apartments, neighborhoods, commercial sectors, and town and city centers.

94. Construction

The following construction investigations and documents will comprise the next phase of study for Triton City:

1. Recommendations of appropriate construction industry(ies) for implementation and (their) capability and support capacities.
2. Alternative construction techniques, with evaluations comparisons, and notes as to the particular appropriateness of each.
3. Performance specifications for construction, materials and equipment.
4. Representative working drawings showing methods and types of materials, fabrication and erection.

A detailed cost analysis will be made and will include estimates for construction and cost-benefit analyses for various alternative proposals.

95. Structural

Efforts in the next stage of research will be given to further examination of each of the topics discussed in paragraphs 52. 2. through 52.8. In depth study will involve determinations of effectiveness, longevity, stability, and capital and maintenance costs.

In addition, systems will be developed for vertical and horizontal fire separation in both the apartment and the core areas. This will include study of materials and con struction evaluations for fire and other safety ratings. Areas where high strength steel and newly developed alloys can be effectively used will be thorcughly explored, as will other modifications to minimize the cost of the basic megastructure frame.

96. Marine

Analysis of the characteristics of waves and the response of the structure to wave action is best carried out by means of model test program in a hydrodynamic test tank, such as the facility at the Hydrodynamics Laboratory of the Massachusetts Institute of Technology. Equipment in the tank can simulate any desired sea state. Shallow water and shore effects can be duplicated with the proper type of "beach" at the end of the tank. The response of the floating body can be determined with the use of a fairly simple model of the structure. This model ehould correctly portray the scale dimensions, overall weight, and approximate distribution of weight. Such a model is normally constructed as a rigid block and does not attempt to scale the elastic qualities of the actual raft and superstructure. The model test program would generate data for comparison with previous experience with rolling of ship structures, and thus, information for final judgment of the suitability of the floating structure for comfortable inhabitance. It would also provide design criteria for wave heights, crest lengths, and other flow characteristics of a particular location.

97. Mechanical

Supplemental study of the mechanical requirements for Triton City will cover the points noted below:

1. Comparative analysis of alternative systems in terms of the following variables:
   a. fixed costs and capital expenditures
   b. operating costs, fuels and maintenance, and amortization of capital
   c. obsolescence and replacement of equipment
2. Preliminary system design, including:
   a. performance specifications
   b. area and volume requirements for both equipment and distribution systems
   c. capacity requirements for each system
   d. plans and sections for the operating system
3. Construction cost estimates

TRITON FOUNDATION, INC. STAFF:

William S. Cope

(Mrs.) Sally W. Fisher

Peter L. Floyd

R. Buckminster Fuller

Walter M. Kroner

Michael Lardner

(Mrs.) Brigit Mathé

Shoji Sadao

David D. Wallace

TRITON FOUNDATION, INC. CONSULTANTS:

Dr. Harvey Evans, Professor, Naval Architecture, M.I.T.

Dr. Ernst Frankel, Professor, Naval Architecture, M.I.T.

Dr. Frank Heger, Simpson, Gumpertz & Heger, Consulting Engineers: Structural

Dr. Melvin R. Levin, Professor, Urban Development, Boston University

Mr. Paul Londe, Londe, Gordon, Parker, Steffen, Consulting Engineers: Mechanical, Electrical, Plumbing, Utilities

Mr. Murray D. Segal, Consultant, Urban Transportation

SUPPLEMENTAL BIBLIOGRAPHY:

1 Abrams, Charles, The City is the Frontier. Harper & Row, New York, 1965.

2 ______________, A.W.W.A. Publications, American Water Works Association, Inc., New York, June, 1967.

3 Baker, Geoffrey and Bruno Funaro, Parking. Reinhold Publishing Corporation, New York, 1958.

4 ______________, Bibliography on Housing, Building and Planning. Housing and Home Finance Agency, Office of International Housing, Washington, D.C., Revised, May, 1964.

5 Chermayeff, Serge and Christopher Alexander, Community and Privacy. Doubleday-Anchor, Garden City, New York, 1963.

6 ______________, "Combating Crime," The Annals of The American Society of Political and Social Science, Vol. 374, November, 1967.

7 ______________, Design in Town and Village. Ministry of Housing and Local Government, Her Majesty's Stationary Office, London, 1958.

8 Doxiades, Costantinos A., Emergence and Growth of an Urban Region: The Developing Urban Detroit Area. The Detroit Edison Company, Detroit, 1966.

9 ______________, The Exploding Metropolis. Doubleday-Anchor, Garden City, New York, 1957.

10 ______________, Flats and Houses: 1958: Design and Economy. Ministry of Housing and Local Government, Her Majesty's Stationary Office, London, 1958.

11 Goodman, Paul and Percival Goodman, Communitas. Random House-Vintage, New York, 1960.

12 ______________, Harlow Expansion Study. Harlow Development Corporation, Harlow, Essex, England, March, 1963.

13 Hatt, Paul K. and Albert J. Reiss, Jr., editors, Cities and Society. The Free Press, Glencoe, Illinois, 1959.

14 ______________, Housing and Planning References. Housing and Home Finance Agency, Office of The Administrator Library, Washington, D.C.; No. 124, November/December, 1964; No. 125, January/February, 1965; No. 126, March/April, 1965.

15 Lynch, Kevin, The Image of the City. The M.I.T. Press, Cambridge, Massachusetts, 1960.

16 Lynch, Kevin, Site Planning. The M.I.T. Press, Cambridge, Mass., 1962

17 Madin, John D. and Partners, Corby New Town Extension Master Plan Report. July, 1965.

18 Mumford, Lewis, The City in History. Harcourt, Brace and World, New York, 1961.

19 ______________, New Communities, A Selected, Annotated Reading List. Housing and Home Finance Agency, Washington, D.C., January, 1965.

20 Thompson, Wilbur R., A Preface to Urban Economics. Publication for Resources for the Future, Inc., The Johns Hopkins Press, Baltimore, Maryland, 1965.

21 ______________, Town Center Report, Part I. Cumbernauld Development Corporation, Chief Architect and Planning Officer's Department, April, 1960.

22 Tunnard, Christopher and Henry H. Reed, American Skyline. The New American Library-Mentor, New York, 1956.

23 Wright, Frank Lloyd, The Living City. The New American Library-Mentor, New York, 1963.

Seaward Ho! -- The Ocean Frontier

by

Adam Starchild

These are ideas that sound like they were snipped from the scenario of one of the new breed of high-tech science-fiction films: A quiet residential neighborhood of homes, shops, and tree-lined open spaces is towed into place and linked up to a town floating off the New Jersey Coast. A community college is lowered onto caissons in the harbor. An entire petrochemical production complex, including apartments and recreational facilities for its workers, an airstrip, and a small shopping area, is erected on a platform floating in the middle of the North Sea. The structural and engineering technique for such grandiose ventures is already a reality. So are the economic, social, and regulatory pressures which may soon make the use of floating platforms and artificial islands a necessity for urban and industrial expansion.

As tinged with the tenor of futurism as all this sounds, the use of water based structures for stationary purposes is an ancient idea. In barely remembered prehistory, neolithic lake dwellers in such diverse locales as Europe, Africa, and the Indian subcontinent built their homes on stilts sunk into the lake bottom. The lake provided protection and defense of the community, a site for the effective disposal of wastes, a mode of transportation, and a source of food. These early design innovators commuted to and from the land, where they hunted and engaged in agrarian pursuits, on wooden rafts which also were used to transport foodstuffs.

What forces are operating in the pre107sent day world to create a twentieth-century version of the stone-age lake-dweller? The attraction the sea offers for industrial expansion is based on the economy and the environmental suitability of ocean-based platforms, coupled with the ability to exploit the oceanic resources in a more efficient manner. Urban expansion into coastal waters is desirable for a different set of factors. The modularity and three-dimensional quality made possible by integrated residential floating structures permits rationalization of many urban functions. It is possible to float on a four-acre platform a structure averaging twenty-stories in height. Design in this three dimensional environment can minimize the spread of high-density urban functions into the countryside and at the same time reduce the absurdity of current urban configurations, which having arisen over the course of centuries are not appropriate to current sociological and technical conditions. A floating neighborhood of 5,000 could maximize the availability of services, minimize fuel consumption, and rationalize transportation, and most importantly, an obsolete or decaying neighborhood could be towed away and replaced in an afternoon's work.

There has been a steady increase in the use of floating platforms for a variety of industrial applications. Since as far back as 1975, Japanese shipbuilding firms have been assembling complete industrial facilities on semi-submersible structures -- facilities which can be transported in one piece directly to the installation site, permitting high quality engineering to progress from conception to operation in a far shorter time than conventional construction for land-based facilities would allow. Platforms up to four acres can be assembled in a conventional shipyard, and engineering of the entire plant, including platform, can proceed as a single unified operation, providing greater design integrity and coherence than would otherwise be possible.

Industrial floating platforms find their prototype in the Mohole Platform, the rig designed to drill through the earth's crust at the so-called Mohovoricic discontinuity in the Pacific. Although the construction of this rig was never completed, the design conception was thoroughly engineered, and its basics were adopted by petroleum concerns in the growing competition to develop offshore oil resources.

In 1975 the Mobil Oil Company began operation of its rig ConDeep, over one hundred miles off the Norwegian coast. The rig, a semi-submersible structure standing on concrete caissons, was equipped with two derricks, a five story apartment complex for the people stationed on the platform, other community facilities for the workers, and a small heliport.

Drilling platforms of various sorts were the first sort of seaborne industrial facility to which attention was paid, but a wide range of other manufacturing and processing ventures are now being designed for such floating enterprises. Kawasaki Heavy Industries, a Japanese firm, in conjunction with an engineering company, built a 2,000 ton-per-day floating desalinization plat for Saudi Arabia.

While industrial platforms are already becoming a reality, the development of platforms for residential purposes has lagged behind. Many designs currently in the works for floating industrial complexes include well-developed units for housing of employees, but, in general, the utilization of platforms for sea based residence in coastal areas is still in the development stage. Not surprisingly, it is among the island-based researchers, such as in Hawaii and Japan, that this research has proceeded with the greatest energy. The Japanese have built what is often considered to be the prototype of the floating city: the 18,000 ton Aquapolis, which was the featured exhibit at the 1975 International Oceans Exposition in Okinawa. Unfortunately, little practical progress has been made since then, despite ever more sophisticated engineering studies of the feasibility of such projects.

Floating platforms are one of the two possible types of marine construction that have been suggested for this sort of seaward expansion. The second type is the artificial island, which is not floated but built up on a "seafill" of solid waste. The advantage of this sort of structure is the utilization of waste material to build up the island, and the simplification of design in terms of assuring maximum stability. Although no high seas artificial islands have as yet been constructed, we shall see in the pages to come that concepts for the construction of such a platform are being developed by a number of concerns. Like the floating platform, the use of building up an artificial land surface in shallow waters is not an exclusive property of twentieth century advanced technology. The "polder" areas in the Netherlands are the results of the use of land fill for centuries to increase the land surface in the coastal area that was available for agriculture.

A variety of forces are conspiring to push us on in our exploration of the sea as a site for residential and industrial construction. As we become more concerned with preserving the environmental quality of our cities, of exploiting the vast resources of the ocean, and of capitalizing on the stability and economic advantage of sea-based industrial facilities, the design and construction of high seas industry will flourish. As we turn our attention to the way in which design considerations are related to the sociological and economic roots of our urban problems, (and need we list them?) the advantages of modular sea-based construction for the expansion of the cities will become evident.

Floating Factories

The day of the floating factory has come. In the field of oil-drilling alone, there are many hundreds of facilities in operation. In addition, there are geophysical drill ships designed for research and exploration, as well as platform-mounted chemical plants, pulp plants, desalinization facilities, and chemical storage facilities. In the design stage are plans for floating a variety of other large scale industrial enterprises. There is a future for seaborne development of oil refineries, electric power generation, both fossil and nuclear, petro-chemical manufacturing, LNG regasification, and nuclear fuel reprocessing. The shipbuilding industry, particularly in Japan, has been shifting gears to address the new demand for seaworthy factories; and some stiff international competition for the lucrative contracts has begun.

The move offshore by large-scale industry is sparked by several concerns. The large manufacturing centers in the United States are located in areas where their further extension and expansion of heavy industry threatens the quality of life for the area's residents. Existing waterways are too small to absorb the load of dumped waste material, and are becoming seriously polluted. Air quality, noise pollution, thermal pollution of waterways, and the propagation of urban sprawl into previously open space are all cited as concerns that industrial expansion onshore must be slowed.

These are valid concerns and, coupled with the growing distrust of nuclear power safety engineering, these worries are giving rise to social and political pressure which makes it increasingly difficult to obtain permission for industrial expansion. The pressures applied by environmental and citizens groups have given rise to regulations at state, federal, and local levels which sharply curtail the construction of new plants. Some of these restrictive regulations are currently being reversed. But the direct pitting of industry against the quality of life is hardly a situation to be courted either by industrialist or environmentalists.

The fact is that if we are to supply the world's people with the goods and services needed for a comfortable style of life and at the same time preserve the environment from degradation, we will need to expend more energy, not less. We are talking about a more sophisticated technology, not a more simplistic one. And we will not be able to do away with the unpleasant fact that there are some vital industrial tasks which should be isolated from the surrounding environment as much as possible but which must go on.

Over eighty percent of our metropolitan areas with a population of a million or more are located on water deep enough to accommodate floating cities, floating communities, and floating factories. In the inshore waters, a variety of facilities can be constructed which need to be close to the port city, which do not present an environmental threat to the landbased community. These facilities include all types of plants needed to directly service the population center: waste treatment, power generation, and deepwater terminals for petroleum, chemicals, liquid natural gas, seafood, and dry bulk substances.

High seas facilities on the other hand would be appropriate for enterprises which do not need to be tied so closely to a port city and which are noxious enough to demand significant isolation from residential locations. Oil refining, nuclear fuel reprocessing, industries related to fishing and deep-water fisheries, and power generation for these plants would be appropriate for high-seas installations.

Environmental Protection

What is the potential for despoiling the vast resource of the ocean through the careless utilization of ocean-based industrial development? Obviously this risk cannot be ignored, for the magnitude of the industrial use of chemically active and radioactive substance is such that it *does* affect the broad ecospheres of the terrestrial environment. The detectable destruction of the ozone through the use of fluorocarbons, the presence of insecticides in fatty tissues of the Antarctic penguins, and a host of other biologic and geophysical horror stories of the last fifteen years attest to this fact. But the fact is that the ocean does have a large capacity to assimilate waste materials. This is due in part to the vast volumetric size of the hydrosphere, and in part to the retardation of chemical processes in the cold and high-pressure environment at the oceans bottom.

The discovery that chemical processes associated with the breaking down of organic wastes are retarded in the deep sea environment was one of those accidental collisions with fact that science is full of. In 1968, the research submarine *Alvin* sank in the Atlantic off of Woods Hole, Massachusetts. The three crewmen escaped, but the ship, with the crew's apples and bologna sandwiches sank to the bottom, a mile below the surface. The vessel was raised and drained ten months later. The sandwiches, the apples, and two thermoses of bouillon were in remarkable condition, despite the fact that they were soaked through with sea water. The apples fared no better than they would have in a normal dry refrigerator, but the rest of the edibles were in a far better state of preservation.

Alvin's sandwiches had spent their ten months under 150 atmospheres of pressure at a temperature of 39° F. Testing the role of these conditions on the chemical degradation of other organic materials, Woods Hole researchers concluded that these substances degraded ten to one hundred time more slowly in the deep sea than under standard temperature and pressure.

The implications of this for the disposal of waste materials in the deep sea are twofold. On the one hand, organic waste dumped in the deep is effectively removed from the ecosystem for a period on the order of a century, limiting the recycling of nutrients. On the other hand, the dispersal of potentially noxious substances from these wastes is also decreased substantially by this process.

In addition, the possibility exists of conscientiously exploiting this slowing down of biochemical processes by dumping wastes in containers made of organic compounds, whose rate of corrosion in the deep sea environment is orders of magnitude slower than had been previously estimated. Thus the dispersal of hazardous waste materials through the hydrosphere would take place far more slowly.

The ocean, if only due to its vast extent, has an ability to absorb environmental insult without significant destruction to its ecological balance. Only small, enclosed outcroppings of the ocean, such as lakes and coves, and the shallow coastal estuaries, are fragile ecosystems. An extensive body of data exists to show that none of the several major assaults on the ocean ecology have had a measurable long term effect; this includes several large oil spills, and the loss at sea of the reactors of at least two nuclear ships. There is little to fear about the introduction of degradable biological waste material into the ocean; in this regard it is useful to remember that there are orders of magnitude more waste material circulated in the oceans due to the fecal material of marine life than there is treated human waste in the system.

Ocean-based industry has other distinct environmental advantages. The ocean can serve as a heat sink for the thermal product of any enterprise. At our current technological level, it would be impossible for us to produce enough heat to alter the thermal equilibrium of the hydrosphere even if we set out specifically to attempt that task. The isolated ocean environment also serves to absorb the noise generated by industrial processes, often a problem in urban factory design. Noise, smog, and visual pollution abatement can all be accomplished by the simple interposition of twenty-five miles of open sea between the plants and the nearest community.

Of course, ocean-based industry cannot go about dumping and polluting in a cavalier fashion, certain that the ocean is an indestructible resource. Although its vast size and biophysical properties give it a great ability to assimilate the by-products of human productive ventures, the ocean, like the atmosphere, will undergo measurable quality deterioration if it is used indiscriminately. For this reason, the National Oceanographic and Atmospheric Administration can set guidelines and regulations regarding the use of specific ocean sites in U.S. waters. To cite but one example, the NOAA ruled a dumping site located 106 miles off the New Jersey Coast as unacceptable due to the possibility that prolonged southerly winds could cause toxic waste materials to back up onto the lower slope or the outer continental shelf and destroy the fishery areas there.

The environmental impact of the construction of an artificial island in the seas is somewhat different from the implication of a large floating platform. Solid waste material from the industrial processes carried out on the island would not have to be dumped into the hydrosphere for disposal, but could be systematically utilized in the expansion of the island. On the other hand, the dredging and filling operations by which such an artificial island would be constructed would destroy the bottom habitat near the island.

There is also the question of the effect an artificial island would produce on the currents between island and shore and the effect of these currents on the shoreline. Of course, an artificial island on the high seas would produce little noticeable effect on the distant shore. But large structures built close to the coastline can alter the process which deposits or erodes sediment from the shoreline. The process through which this occurs can be understood in terms of an optical model of the wave propagation in the sea. Waves will be refracted and diffracted around the edge of the structure, resulting in an area of low wave energy, and there is formed a "shadow" behind the island. If this shadow reaches the shore, there will be an increase in the deposition of sand in this region. In turn, this deposition will block the component of sand transport parallel to the shore, resulting in erosion downshore from the shadow.

Thus we see that while the offshore development of industry on artificial islands or floating platforms cannot be viewed as a panacea for environmental concerns which permit us to industrialize and to discard wastes without consideration of impact on the ecosystem, there are significant environmental benefits from the placement of some industrial enterprises offshore.

Design and Engineering

Two distinct technologies have been proposed for the offshore development of industry; the artificial island, and the floating platform. The two technologies produce offshore structures with very different economic and social characteristics which stem from differences in assembly, modification, location, and so on.

Floating platforms are generally structures of the "semi-submersible" type. This term is used to indicate that there is a significant part of the structure located beneath the surface of the sea. The bulk of the structure's mass, providing the platform's buoyant properties and inertial stabilization, are located not only below the surface, but well below the zone of turbulence and wave action characteristic of the air-ocean interface.

Only thin columnar structures, built on the deep submerged base, penetrate the zone of turbulence. These thin structures provide little cross section for the absorption of the energy of surface turbulence. This minimizes any mobility of the structure caused by wave motion at the ocean surface. The platform itself, which can be of a size up to four acres in area with current engineering techniques, rests upon columns or pylons well above the maximum height for waves in the area.

The resulting structure is remarkably stable. It is possible to design a semi-submersible platform such that unmoored in the high seas it will never experience an acceleration of .02g or greater. This figure represents the limit of human ability to detect motion. It is also well below the level of acceleration that many urban codes require in structures which must be safe in time of earthquake or land tremor.

The floating platform also has specific economic advantages. While plants assembled on an artificial island must be built by standard landbased building techniques, the integrated prefabrication of plant and platform that the semi-submersible permits brings about significant cost advantages. A Westinghouse-Tenneco study of the feasibility of a floating nuclear facility pointed to a list of specific economic advantages: The elimination of site-specific design and engineering, the elimination of earthquake precautions, shipyard fabrication, supply by barge rather than truck. In addition, the seaborne nature of the plant means that there is no need to prepare the foundation, or to build roads, or other access routes such as rail lines to the new facility.

The conclusion of many cost analysis studies on the feasibility of floating plant construction indicated that where the size of the structure is large but the volume to area ratio small, the floating platform is an appropriate form. Thus factories and office construction is reasonable on a semi-submersible structure, while an airfield would be more economically constructed offshore on the artificial island type of platform.

The offshore artificial island lacks some of the specific advantages of the floating platform. It is not immune to tremor and earthquake, and is prey to the same forces that act on natural coastal structures: wind and water will tend to erode the fill out of which the island is constructed if armor is not placed to serve as a sea wall around the island. The shape of the island also plays a role in the erosion-resistance of the structure. An elliptic island placed with the long axis parallel to the strongest current will erode significantly more slowly than a circular island.

The island also presents us with a risk of flooding which the floating platform, elevated above the waves and immune to tidal action, did not. Sea defenses including the construction of sand beaches on the island and sea walls have been proposed to eliminate the possibility of flooding.

As mentioned previously, the specific advantages cited for the construction of artificial island over floating platforms is in platform area. For purposes requiring a large area for a relatively small mass of actual plant -- that is, for low density activity, the island is a more cost effective venture than the platform. The direct incorporation of processed waste material into land fill for island building and expansion is a second important feature in the overall cost picture of the artificial island approach to seaward expansion.

Sea islands have been proposed as an ideal location for centralized waste processing for a variety of other reasons. Centralized processing of solid waste permits the storage and combustion of waste in an environment which will not need to be protected from the noise and emissions of such a venture. It also permits secondary industries such as fertilizers and others which utilize the byproducts of waste disposal processing. Moreover, the cost per unit of waste processing is extremely high in small industries; no doubt there is a connection between this fact and the tendency of many small concerns to attempt often-times dangerous circumvention of laws governing waste disposal.

The proposed component parts, then, of an island centralized waste processing complex, would include a power plant, an incineration plant, a chemical treatment facility, and a scrap processing plant. In addition, such ancillary functions as an air separating plant, a desalinization facility, and a fertilizer plant.

Floating factories and factories on large manmade islands may play an important role in the industry of the future. Such seaward expansion of our industrial capabilities gives us the opportunity to continue to develop industrial solutions to the constant problem of human comfort, while minimizing the negative impact industrial complexes often have on the physical and biological environment. Floating factories are already a reality. Hundreds of semi-submersible oil-rigs are drilling for petroleum in the world's offshore oilfields, and orders are coming from all parts of the globe for modular plant facilities that can be anchored offshore. The unitary construction of these factories permits highly cost effective construction for high-density applications. At the same time, technologies are being developed for large artificial islands which have many advantages for the offshore development of low density commercial applications, such as airfields, and waste disposal.

INTO THE FUTURE . . .

Men go to war to gain access to the ocean. Since the daybreak of human knowledge, the ocean has played a vital role in the life of mankind. The ocean was a harsh and terrifying force which could take a life and close over the dead with hardly a ruffle; yet it was a source of food, a means of transportation, a protective natural barrier, a battlefield.

As science and technology have advanced over the past centuries, man's relationship to the world's oceans have changed. The frail craft which a few hundred years ago crisscrossed the oceans blown by the winds have been replaced with nuclear powered submarines, automated containerships, and supertankers. The nets and lines of fishermen are now guided and controlled by shipboard computers, remote electronic sensing, and satellite communications. Since Neolithic times men have built homes above the water for protection from animals and enemies; today floating hotels house thousands of offshore oil field workers.

Man now approaches the traditional seafaring activities with the tools and techniques of the late twentieth century. But new uses of the ocean have emerged as well over the last few decades. The discovery of rich oilfields in the subsoil of the continental shelf lands has created an entire industry. The development of technology to access the manganese nodules which carpet so much of the deep sea bed promises vast mineral wealth. The purposeful cultivation of aquatic plant and animal life has created an aquaculture which hints at becoming as important a source of food as fishing over the next fifty years.

The advance of technology has reshaped our vision of the oceans. The changing relationships among nations have also helped to form the role of the seas in human life. Since the Pope drew the line of demarcation in the fifteenth century to separate the Spanish and Portuguese colonial domains, sea law has reflected the actual political, economic, and military relationships among nations. From the doctrine of the freedom of the high seas, which formed the basis of international conventions on the sea for so long, to the modern movement to enclose even larger portions of the ocean within the bounds of national sovereignty, the utilization of the sea and its resources has depended in large measure on the international political climate as it affects the boundaryless ocean.

None of the resources of our planet offered up to humanity's voracious appetites are inexhaustible; this is the great lesson of the nineteen hundreds. The ocean promises hoards of minerals, a bountiful supply of foodstuffs, and millions of barrels of fuel oil, at a time when he scarcity of further untapped resources on the land areas is becoming ominous. But the oceanic resources have their limits as well. The wealth of the world ocean system can be polluted and depleted just as the land based resources have been. Yet the oceans also hold out an exciting challenge, for we are beginning our exploitation of the ocean's fabulous riches with a consciousness of the need for conservation and environmental protection. If this consciousness is maintained at every step, we face the exciting prospect of a truly rational and sound program of oceanic development, in which the greatest amount of use is made of the oceans without damaging the ocean system appreciably.

The excitement over the new possibilities of man's relationship to the ocean, the rapid growth of our knowledge of the ocean and the resources it holds, has made oceanography one of the most important sciences of the era. The number of oceanographers is growing faster than the size of any other aspect of the scientific community. Developments in oceanography have led to new insights in geology, biology, and even in the understanding of the properties of water itself. From submarine expeditions to the bottom of the Marianas Trench to experimental turtle farms, oceans scientists have been pushing back the frontier of knowledge about the oceans. The understanding of the spreading of the sea floor alone has revolutionized our notion of the history of our planet. The theory of continental drift, proposed originally to account for the similarities in the coastal configurations of Africa and South America, is now an accepted part of geological knowledge.

There is much excitement over the newly found resources within the sea. Perhaps the single most important contribution this oceanic wealth can make to the well being of the human race is in he area of food. We live on a planet where the majority of nations are inadequately fed. The sea promises to serve as a source of food for millions of people who are currently suffering from caloric or protein insufficiencies. Of course, the inequitable distribution of food and wealth between the rich and poor nations contributes to this shortage and needs to be directly addressed, as many concerned with the distribution of the sea's wealth have pointed out, but the food resources within the sea cannot but help to alleviate the problems of world hunger. Many of the plants constituting the bulk of the diet in some of the less developed countries contain on the order of one-tenth the protein per calorie of fish.

The sea offers a vast supply of food. Conventional fishing, aided with modern remote sensing devices can continue to play a significant role in this process. New technologies can augment it. The harvest of seaweed and plankton promise to create a valuable new source of nutrition; the plankton krill is already available in Soviet markets. Aquaculture and mariculture as well are becoming increasingly important in projections of the ocean food resources. Legislation governing many aspects of aquaculture development has been passed by congress, reflecting the seriousness with which these new food sources are being taken by policy makers.

The signs on the highway read "Food and Fuel." Certainly these are the two greatest individual appetites of modern society. Living in a nation which is the world's leading food exporter, we are often more aware of the scarcity of food. The ocean promises to provide us with a bounty of petroleum as well as a rich harvest of edibles. The existence of oil on the continental shelf has been known since the end of the last century, as we have learned earlier, but it is only in the last two decades that offshore recovery of oil has been a major industry.

Immediately as we begin to increase the mineral and edible wealth we extract from the sea we are forced to confront the issues of environment control, conservation, and pollution. A major controversy rages between those who would preserve the rich Georges Bank area off the coast of Newfoundland as a fishery area, and those who would prefer to see it developed as a source of petroleum. A single spill or blowout from a well in the Georges Bank could indefinitely destroy its value as a fishing ground.

But even short of these major ecological catastrophes, questions of conservation and protection are raised. Overfishing is common in many of the richest fishing grounds, with the result that stocks are beginning to show signs of depletion. Similarly, the routine use of the seas, not catastrophic spills or well accidents, contribute by far the largest share of oceanic pollution. Thor Heyerdahl, the ocean-going adventurer, says that he and his crew mates saw oil slicks in the ocean on 43 of their 49 days at sea in the reed and papyrus boat, *Ra*. If the use of the sea for food and fuel is to be increased, stern measures will have to be taken to prevent further fouling of the ocean system.

No doubt there is an aspect of this problem which is intensely political, involving as it does the shared commitment of a plenitude of nations to the necessity of preserving the ocean environment. The conflicts among nations has ever played a great role in the relationship between man and the sea; the sea has as often served as a battleground as it has a source of food and other vital resources.

The conflicts were between the rich nations and the poor, rather than between east and west. They erupted most forcefully over the marine mineral resources of the deep sea bed. The technology currently exists to extract a fantastic wealth of minerals from the sea in the form of ferromanganese nodules, but the disagreements as to the extent to which these resources are to become appropriated by the countries with the existing means to extract them, in contradiction to the UN's doctrine of the ocean being "the common heritage of mankind," has precluded all development of these resources. So the development of the sea has moved forward slowly when at all, with the less developed nations pressing for a single world enterprise to control the mining of the seabed, and the powerful mining corporations of the developed world pushing instead towards the pole of complete freedom to mine in the high seas.

It is not only the question of mineral rights and mine development which has required political response. The growth in the demand for fish has raised the competitive stakes in world fishing and led directly to the claiming of 200-mile fishery zones by most major fishing states, a move which frightens shipping and military interests concerned with the protection of the right to transit.

Almost thirty years ago Senator Clairborne Pell, who has long been a leading voice on Capitol Hill warning of the dangers and promise the ocean holds, wrote these words:

After millenniums of exploiting and often destroying the riches of the land, man is now hovering acquisitively over the wealth of the oceans that cover three quarters of the earth. In the no-man's-land of the sea bed, a scramble for minerals and oil, for new underwater empires secured by the advancing armies of technology, could well set a new and wider stage of world conflict. (Saturday Review, Oct. 11, 1969)

Now, many years later, the words of Senator Pell still ring true. While the possible resources the ocean offers us seem larger with each new study, with every oceanic survey, little progress has been made in the political arena. Meanwhile, the destruction of the hydrosphere continues. Oil continues to spill from a blowout of an offshore well. Chemicals and pesticides dumped in rivers and canals continue to find their way into the ocean. Conservationists look in fear and trembling at the repercussions of offshore petroleum development in the Gorges Bank fishing grounds.

But our use of the oceans is only in its initial phase, and we approach it with a far deeper insight into the fragility of the earth which appears to be so bountiful.

What humanity has learned about the delicate nature of the earth's ecological balance in the past twenty years brings about a natural sense of caution in undertaking a new venture, especially one which touches on the ocean system, which was the mother of life and remains essential for life on the planet. Similarly, it is only prudent to bear in mind the critical role played by international agreement and disagreement in any enterprise based on the high seas.

But there is no reason for turning away from the prospect of an entirely new degree of magnitude in man's usage of the seas. These ancillary problems are merely a part of the challenge of learning to live with our planet and each other. For the fact is that the development of the oceans has been relatively unattended by disasters. The massive Santa Barbara oil spill which justly caused such a public outcry in 1974 did not turn out to be merely the first in a long string of horror tales. Regulations brought about by the public's concern about the environment have led to relative safety in offshore petroleum development. Routine operations by vessels are far more dangerous to the ocean environment than spills from offshore wells. Similarly, new developments in fish farming and stocking, new conservation measures, and the development of new species as fishery stock have enabled the fishing industry to increase production without endangering the basic food fish.

Without tapping the resources of the ocean environment we cannot continue the economic growth to which we have become accustomed. Yet, to damage the ocean environment is to risk human life entirely. Between these two possibilities lie a multitude of ethical considerations, economic tradeoffs, technological advances, and political conflicts. The future of the oceans is truly the future of humanity.

As pragmatic as such a program may appear, its most important and long-reaching benefits will surely be sociopsychological rather than simply economic. Along with the development of an ocean resource program, there will be an attendant need for education and training, as well as the creation of new occupational fields in diving, boating, underwater exploration and construction, etc. The cadre of leaders that could evolve out of these occupations would be unique in the world; they would be the ocean pioneers, capable of spearheading a movement to excite the imagination of the island youth and develop in the population a pride and esprit de corps that would be truly an island accomplishment.

When in history has a people who have staked out a claim or wrested a living from the virgin wilderness feared the future or become an indolent or dependent people? The answer is never: It is only when there are no frontiers to cross that initiative and the courage to face real choices vanishes.

About the Author

Adam Starchild is the author of numerous business books (including *The Ocean Frontier*, published by University Press of the Pacific), editor of an anthology of Russian science fiction, a member of the Extropy Institute, and a life member of the Lighter-Than-Air Society, the World Future Society and the Libertarian Futurist Society. His personal website is at http://www.adamstarchild.com

CPSIA information can be obtained
at www.ICGtesting.com
Printed in the USA
BVHW01s1103151018
530210BV00003B/241/P